Dr. Luis Glaser
Department of Biochemistry
Washington University
School of Medicine

GROWTH CONTROL IN CELL CULTURES

Symposium on

GROWTH CONTROL
IN CELL CULTURES (1970 : London)

A Ciba Foundation Symposium

Edited by
G. E. W. WOLSTENHOLME
and
JULIE KNIGHT

CHURCHILL LIVINGSTONE
Edinburgh and London
1971

Printed in Great Britain

© Longman Group Ltd. 1971

First published 1971
Containing 74 illustrations

I.S.B.N. 0 7000 1517 5

QH
585
.593
1970

Contents

Contributors

Symposium on Growth Control in Cell Cultures, held 14th–16th October, 1970

M. G. P. Stoker (Chairman)	Imperial Cancer Research Fund, Lincoln's Inn Fields, London WC2A 3PX, England
M. Abercrombie	Strangeways Research Laboratory, Worts Causeway, Cambridge, England
J. B. L. Bard	MRC Clinical and Population Cytogenetics Research Unit, Western General Hospital, Crewe Road, Edinburgh 4
F. Bergel	Magnolia Cottage, Bel Royal, Jersey, C.I.
M. M. Burger	Program in Biochemical Sciences, Moffett Laboratories, Princeton University, Princeton, New Jersey 08540, U.S.A.
R. R. Bürk	Imperial Cancer Research Fund, Lincoln's Inn Fields, London WC2A 3PX, England
G. D. Clarke	Imperial Cancer Research Fund, Lincoln's Inn Fields, London WC2A 3PX, England
D. D. Cunningham	Curriculum of Environmental Interactions, California College of Medicine, University of California, Irvine, California 91664, U.S.A.
R. Dulbecco	The Salk Institute for Biological Studies, P.O. Box 1809, San Diego, California 92112, U.S.A.
T. R. Elsdale	MRC Clinical and Population Cytogenetics Research Unit, Western General Hospital, Crewe Road, Edinburgh 4
I. B. Holland	Dept of Genetics, The University, Leicester LE1 7RH, England
R. W. Holley	The Salk Institute for Biological Studies, P.O. Box 1809, San Diego, California 92112, U.S.A.
I. A. Macpherson	Imperial Cancer Research Fund, Lincoln's Inn Fields, London WC2A 3PX, England
N. A. Mitchison*	National Institute for Medical Research, Mill Hill, London N.W.7, England
L. Montagnier	Institut du Radium, Faculté des Sciences, Bâtiment 110, 91 Orsay, France
J. D. Pitts	Dept of Biochemistry, University of Glasgow, Glasgow W.2, Scotland
G. Pontecorvo	Imperial Cancer Research Fund, Lincoln's Inn Fields, London WC2A 3PX, England

* *From September 1971*: Dept of Zoology, University College, Gower St, London W.C.1.

CONTRIBUTORS

H. Rubin — Virus Laboratory, University of California, Berkeley, California 94720, U.S.A.

M. Shodell — Imperial Cancer Research Fund, Lincoln's Inn Fields, London WC2A 3PX, England

Joyce Taylor-Papadimitriou — Dept of Virology, Theagenion Cancer Institute, Thessaloniki, Greece

G. J. Todaro — Viral Carcinogenesis Branch, National Cancer Institute, National Institutes of Health, Bethesda, Maryland 20014, U.S.A.

L. Warren — Dept of Therapeutic Research, School of Medicine, University of Pennsylvania, Philadelphia, Pa., U.S.A.

R. A. Weiss — Dept of Anatomy, University College, Gower Street, London W.C.1, England

The Ciba Foundation

The Ciba Foundation was opened in 1949 to promote international cooperation in medical and chemical research. It owes its existence to the generosity of CIBA Ltd., Basle (now CIBA-GEIGY Ltd.), who, recognizing the obstacles to scientific communication created by war, man's natural secretiveness, disciplinary divisions, academic prejudices, distance, and differences of language, decided to set up a philanthropic institution whose aim would be to overcome such barriers. London was chosen as its site for reasons dictated by the special advantages of English charitable trust law (ensuring the independence of its actions), as well as those of language and geography.

The Foundation's house at 41 Portland Place, London, has become well known to workers in many fields of science. Every year the Foundation organizes six to ten three-day symposia and three or four shorter study groups, all of which are published in book form. Many other scientific meetings are held, organized either by the Foundation or by other groups in need of a meeting place. Accommodation is also provided for scientists visiting London, whether or not they are attending a meeting in the house. The Foundation's many activities are controlled by a small group of distinguished trustees. Within the general framework of biological science, interpreted in its broadest sense, these activities are well summed up by the motto of the Ciba Foundation: *Conscient Gentes*—let the peoples come together.

CHAIRMAN'S INTRODUCTION

M. G. P. STOKER

IT is nearly two years since Professor Franz Bergel initiated a symposium here on homeostatic mechanisms (*Homeostatic Regulators* 1969), and although we didn't come to any clear views about the principles involved, and certainly not the mechanisms, it was an extremely stimulating meeting and opened up a number of facets and views and a very wide-ranging discussion. After it, some of us felt that it would next be useful to have a meeting about regulation in the more easily controlled cell culture systems. Unfortunately, there is so far no such thing as real homeostasis in any cell culture system; in effect we give cells all the substances we believe they need and then ask naively why they sometimes fail to grow. In other words, we study restriction of growth rather than homeostasis, but it is the nearest we can get at present in cell culture systems that can be readily manipulated.

Interest in restriction of growth in cultured cells was greatly stimulated by the discovery of viral transformation *in vitro*, which removes many of the restrictions. The dramatic change in the growth capacity of virally transformed cells led very quickly to simple assumptions about cell contact as a regulator of growth, despite continual protests from Professor Michael Abercrombie that "contact inhibition" referred to movement and not to growth. The role of cell contact in growth regulation has been in and out of favour ever since, generally alternating with serum factors, and my own approach and interpretation have fluctuated as much as anyone's. During the last ten years or so of study a lot of knowledge has accumulated, and it has led to more meaningful experiments, but there has been little progress towards an understanding of the situation, except perhaps in the last year or so.

In this meeting, we shall be discussing new data and we shall try to formulate theories about mechanisms. I hope that at least we shall end up with some good ideas about the experiments to do next.

REFERENCE

Ciba Foundation Symposium (1969) *Homeostatic Regulators*, London: Churchill.

STUDIES OF SERUM FACTORS REQUIRED BY 3T3 AND SV3T3 CELLS

ROBERT W. HOLLEY AND JOSEPHINE A. KIERNAN

*The Armand Hammer Center for Cancer Biology, The Salk Institute for Biological Studies,
San Diego, California*

Our studies of certain mammalian cell growth factors began as an investigation of "contact inhibition" of cell division in 3T3 cells (a cell line derived from mouse embryos). Under normal culture conditions, 3T3 cells stop growing after they have formed a monolayer, which has approximately 3×10^4 cells per cm^2. The work of Todaro, Lazar and Green (1965) had shown that the addition of serum to a monolayer of resting 3T3 cells leads to initiation of DNA synthesis in some of the cells. We looked into this effect of serum since it seemed that this might give some clue to the nature of "contact inhibition" of cell division. In our studies we were impressed by the fact that the cell density at which "contact inhibition" takes place depends on the amount of serum added to the culture medium (Holley and Kiernan 1968). This is shown in Fig. $1a$.

The interaction between serum concentration and cell density of 3T3 cells is shown in a different way in Fig. $1b$. Here a "steady-state" concentration of serum is maintained by growing the cells on a small cover-slip in a large dish, with daily fluid changes. Again the final cell density is dependent on serum but a considerably higher cell density is obtained at a given concentration of serum when the cells are not allowed to deplete the medium.

The results shown in Figs. $1a$ and b are consistent with each other, since assay of depleted 10 per cent serum medium after 3T3 cells have stopped growing—that is, the conditions of Fig. $1a$—indicates that the depleted medium still contains growth activity for 3T3 cells equivalent to 1–2 per cent serum. This is approximately the "steady-state" concentration of serum that gives the same cell density (Fig. $1b$). Depletion of the medium by growing or resting 3T3 cells poses many interesting questions, but definitive studies of depletion are complicated by the fact that there seem to be several serum factors, as is discussed below.

The virally transformed SV3T3 cell does not exhibit "contact inhibi-tion" of cell division. An explanation for this is furnished by the fact that SV3T3 cells have a very low serum requirement for growth. At very low serum concentrations the growth of SV3T3 cells is limited by serum (Fig. 1a). (That the low number of SV3T3 cells found at low serum concentrations is not due to loss of cells from the surface but is due to a lengthened generation time has been confirmed by Ingrid Klinger in this laboratory by time-lapse photography.) The serum factor that is required by SV3T3 cells is not the same factor that is most limiting of the

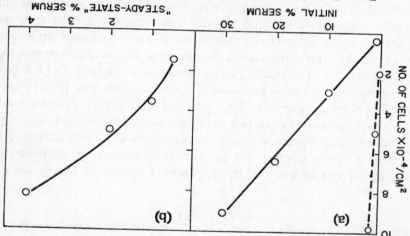

Fig. 1. Dependence of cell density on serum concentration. In Fig. 1a (left), the cell density is that attained after growth had largely ceased (4 days) in media with the serum concentration shown. ———, 3T3 cells; – – –, SV3T3 cells. In Fig. 1b (right), the cell density is that attained by 3T3 cells on a 12-mm coverslip after growth had ceased (5 days) in 10 ml medium, fluid being changed daily, with the serum concentration shown.

growth of 3T3 cells, since 10 per cent serum medium depleted of growth factor by 3T3 cells still assays as 10 per cent serum with SV3T3 cells, although it assays as only 1–2 per cent serum with 3T3 cells.

The observation that the cell density attained before 3T3 cells become "contact inhibited" varies with the amount of serum can, of course, be interpreted in various ways. One interpretation is that contact between cells is of primary importance in inhibiting growth and the effect of serum is secondary; in some way serum "antagonizes" contact between the cells. Alternatively it could be that the growth of 3T3 cells is controlled primarily by growth factors in the serum, and contact and crowding affect growth by limiting the uptake or utilization of the serum factors.

Although these two interpretations are in some respects operationally indistinguishable, we have chosen the interpretation that emphasizes serum factors because it is the simpler hypothesis and it encourages a direct experimental attack—the isolation and study of the serum factors.

Much of our effort has therefore been devoted to the fractionation of calf serum and rat serum with the aim of isolating purified growth factors. Rat serum has been used in many experiments because its greater activity (approximately 2·5 times the activity of calf serum) makes it easier to assay fractions.

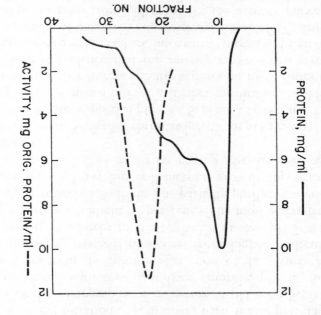

Fig. 2. Chromatography of rat serum, pH 4, on Sephadex G100 at pH 4 in 0·01 N-sodium chloride, pH 4. ———, protein; – – –, activity for 3T3 cells.

As might have been expected from the experience of others, isolation of a highly purified growth factor from serum has encountered technical difficulties. Activity is lost readily and the activity that remains often appears in various fractions. For example, chromatography of serum on DEAE-cellulose columns gives poor recovery of activity and the activity that is recovered is spread across most of the fractions. Also it seems that after many fractionation procedures combinations of fractions are somewhat more active than individual fractions. For example, ammonium sulphate gives some fractionation but the fractions seem to give better growth when they are recombined. Frequently none of the fractions obtained has a higher specific activity than the starting serum, because of losses of activity.

Recently we have avoided some of these difficulties by working with serum at low pH. Much of the activity of serum survives pH 3 or pH 2. In agreement with the findings of Tritsch, Bell and Grahl-Nielsen (1968) for other serum growth factors, 3T3 and SV3T3 growth activity is largely removed from serum by charcoal at pH 3 but not at pH 7. This suggested that the growth factors might be dissociating from carriers at low pH, and, therefore, gel filtration of serum was tried at low pH. Fig. 2 shows the results of chromatography of rat serum at pH 4 on Sephadex G100 packed and eluted with 0.01 n-sodium chloride, pH 4. Activity for 3T3 cells elutes from the column after most of the protein has eluted. Whether the active material dissociates from higher molecular weight material or simply adsorbs to the Sephadex at low pH is not known. Recovery of activity is about 60 per cent. The most active fractions are purified approximately five-fold. The partially purified material will not replace serum completely but will increase growth approximately five times when it is added to media containing low (0.6 per cent) serum concentrations.

Fractions from the Sephadex chromatography at pH 4 can be lyophilized and then rechromatographed at pH 3 on Sephadex G100, in 0.1 n-sodium chloride, pH 3. As shown in Fig. 3, further purification is obtained, and the best fractions now have a specific activity 20 to 50 times that of rat serum. There is evidence that the activity curve actually represents a combination of two active factors, since simultaneous addition of fractions from the two sides of the activity curve gives growth of 3T3 cells to higher densities than fractions from either side of the activity curve alone. A serious difficulty with this partially purified preparation is the apparent lack of stability when it is neutralized. This has limited studies of further purification. The cause of the instability is unknown.

On the Sephadex G100 columns, activity for SV3T3 cells separates from activity for 3T3 cells. Paul, Lipton and Klinger (1971) in this laboratory have found that the best separation takes place at pH 2, with SV3T3 activity appearing in two peaks, one before and one after the 3T3 activity peak. These workers also found that activities for the two cell lines can be separated partially by electrophoresis of rat serum, for example, on Pevikon at pH 7.6 in Veronal buffer.

Our most promising results have been obtained very recently with commercial bovine serum fractions. Chromatography of commercial β-globulin (Cohn fraction III) at pH 3 on Sephadex G100 in 0.1 n-sodium

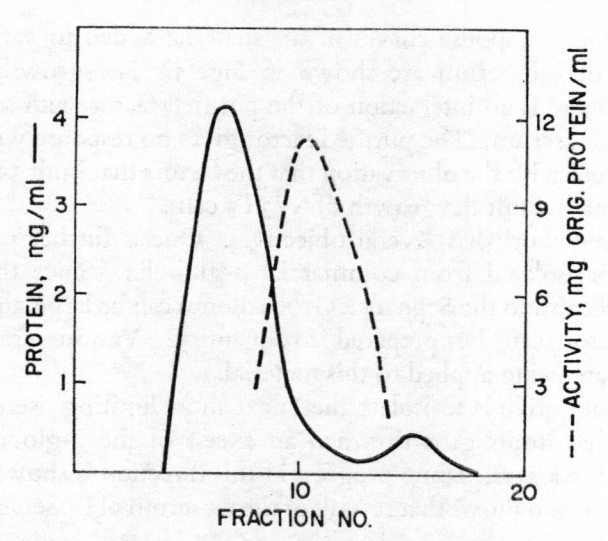

Fig. 3. Rechromatography of lyophilized fractions 25–30 from Fig. 2 at pH 3 on Sephadex G100 in 0·1 N-sodium chloride, pH 3. ———, protein; – – –, activity for 3T3 cells.

chloride, pH 3 gives the results shown in Fig. 4. Activity separates well from most of the protein. The best fractions have specific activities a few hundred times the activity of calf serum. The addition of 1 μg of this material per ml of medium gives a significant growth response with

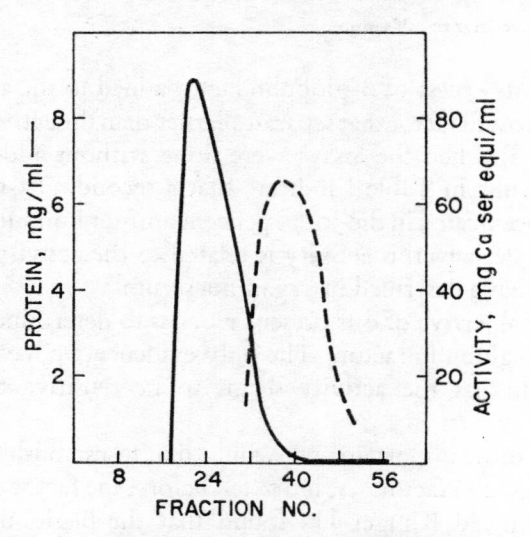

Fig. 4. Chromatography of commercial bovine β-globulin (Cohn fraction III) at pH 3 on Sephadex G100 in 0·01 N-sodium chloride, pH 3. ———, protein; – – –, activity for 3T3 cells.

3T3 cells. Dose–response curves of the material added to varying concentrations of calf serum are shown in Fig. 5. The growth response shows that there is an interaction of the purified factor with some other factor(s) in calf serum. The purified factor gives no response with SV3T3 cells, consistent with the observation that the factors that limit 3T3 growth do not normally limit the growth of SV3T3 cells.

Our present work has several objectives. One is further purification of the factor isolated from commercial β-globulin. Since the starting material is cheap and the Sephadex G100 column can be large, the partially purified factor can be prepared in quantity. Various fractionation procedures are being applied to this material.

Another objective is to isolate the "next most limiting" serum factor, the factor that limits growth when an excess of the β-globulin factor is added to the assays. Some progress in this direction is shown in Fig. 6 and Table I. Fig. 6 shows that reassay of the rat serum pH 4 Sephadex G100

TABLE I

ASSAY OF 3T3 GROWTH ACTIVITY IN AMMONIUM SULPHATE FRACTIONS OF CALF SERUM

Ammonium sulphate fraction	Protein concentration, mg/ml	Specific activity,* assayed alone	Specific activity,* assayed with added β-globulin factor
0–35%	48	0.2	0.6
35–50%	46	0.6	2
50–70%	18	0.4	4

* Activity per mg, relative to calf serum.

fractions with an excess of β-globulin factor added to the assays discloses an additional growth factor that separates earlier than the activity previously observed (Fig. 2) when the assays were done without added β-globulin factor. The results in Table I indicate that a second 3T3 growth factor seems to be concentrated in the 50–70 per cent ammonium sulphate fraction of calf serum. Perhaps this activity is related to the activity Jainchill and Todaro (1970) have described in "agamma serum".

Still another objective of our present work is to determine the chemical nature of the β-globulin factor. The only evidence we have at present is the observation that the activity seems to be sensitive to proteolytic enzymes.

Finally, and most important, we would like to establish the biological role of the β-globulin factor. As indicated before, the factor cannot replace serum. Also, Ingrid Klinger has found that the β-globulin factor has little or no activity in stimulating "crawling" of 3T3 cells from the edge of a wound; the "crawling" phenomenon must therefore be dependent

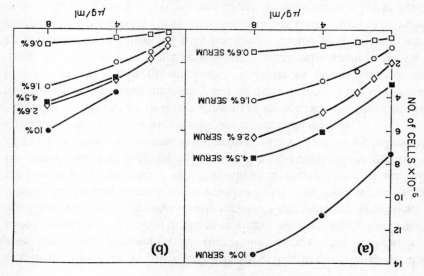

FIG. 5. Dependence of growth of 3T3 cells, set up in varying concentrations of serum, on the addition of varying amounts of β-globulin fraction. Fig. 5a (left) shows final numbers of cells. Fig. 5b (right) shows the numbers of cells above the controls; that is, the numbers in Fig. 5a minus the numbers attained in the varying serum concentrations without added β-globulin fraction.

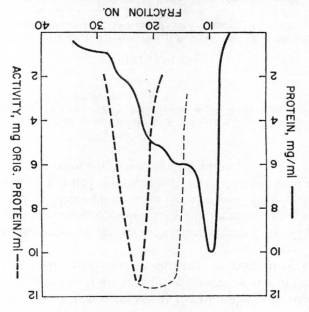

FIG. 6. Reassay of fractions from the column shown in Fig. 2 with β-globulin fraction added to the assays, – – –.

at least in large part on other serum factors. The β-globulin factor does give significant stimulation of thymidine incorporation when it is added to resting 3T3 cells in depleted medium, but the stimulation is much less than that given by whole calf serum. Our working hypothesis is that the β-globulin factor has some action in G1, but other serum factors are also required to initiate DNA synthesis in the majority of the cells. Because of the fact that the β-globulin factor stimulates the growth of 3T3 cells but does not stimulate the growth of SV3T3 cells we believe the biological role of this factor may be of considerable interest.

Finally, as a footnote to this discussion of serum factors, it is of interest that the effects of serum deficiency can be quite delayed. We have found that mitotic 3T3 cells obtained after 8 hours of serum "step-down" (from 10 to 0.4 per cent serum), when subsequently grown in 10 per cent serum, show a 4 to 5 hour delay in the next mitosis, when compared with mitotic cells isolated from 10 per cent serum. The results suggest that the series of events that initiate mitosis actually begin before the previous mitosis.

SUMMARY

For 3T3 cells, the "saturation density" observed is an approximate measure of the serum concentration at the time the cells stop growing. SV3T3 cells require much less serum for growth. Growth rates of both 3T3 and SV3T3 cells are dependent on serum concentration at low levels of serum.

Partial fractionation of the serum factors required by 3T3 cells and by SV3T3 cells has been accomplished by gel filtration on Sephadex G100 at low pH and by electrophoresis on Pevikon or Cellogel. Activity for 3T3 cells has been purified approximately 40-fold from serum and several hundred fold from commercial bovine β-globulin.

Acknowledgements

This work was supported in part by the U.S. National Cancer Institute, the National Science Foundation, and the Elsa U. Pardee Foundation. R. W. H. is an American Cancer Society Professor of Molecular Biology.

REFERENCES

HOLLEY, R. W. and KIERNAN, J. A. (1968) *Proc. natn. Acad. Sci. U.S.A.* **60**, 300–304.
JAINCHILL, J. L. and TODARO, G. J. (1970) *Expl Cell Res.* **59**, 137–146.
PAUL, D., LIPTON, A. and KLINGER, I. (1971) *Proc. natn. Acad. Sci. U.S.A.* **68**, 645–648.
TODARO, G. J., LAZAR, G. K. and GREEN, H. (1965) *J. cell. comp. Physiol.* **66**, 325–333.
TRITSCH, G. L., BELL, J. A. and GRAHL-NIELSEN, G. (1968) *Nature, Lond.* **219**, 300.

DISCUSSION

Shodell: Is it general in your system that the final density the cells achieve is correlated with their initial growth rate?

Holley: The initial growth rate is constant above 4 per cent serum in the medium; it is constant at higher serum concentrations with added β-globulin fraction, but the final cell density depends on the amount of serum or added β-globulin fraction.

Shodell: With less than 4 per cent serum, when you are getting an increase in initial growth rates with increasing serum, will an increase in the amount of the β-globulin fraction give a concomitant increase in the final cell density achieved?

Holley: They both increase, but I don't know the exact relationship.

Dulbecco: There is evidence that serum does many different things and some of them can be separated out by proper assay procedures, such as stimulation of DNA synthesis in a wound or cell layer, or the survival of cells; and they might be independent effects of serum. It would be worthwhile improving the assay procedure to test for some of these effects separately rather than for "growth", which probably detects a summation of effects.

Holley: We try to do this. The β-globulin fraction will not lead to survival by itself, when simply added to medium without serum. We have studied the incorporation of thymidine in a non-synchronous cell population, where one is measuring a change in growth rate. The β-globulin fraction stimulates incorporation. If we add the β-globulin fraction to resting 3T3 cells in G1, the way Dr Bürk does his assays, the cells are stimulated to go into S phase, but less well than with serum. There is evidently interaction with something else in serum.

Stoker: Does charcoal-depleted serum allow survival?

Holley: Yes.

Cunningham: Lieberman's group has shown that injection of Cohn fraction III into rats stimulates DNA synthesis and mitosis in the liver (Short *et al.* 1969). They found that fraction IV alone was inactive, but that it enhanced the effect of fraction III by two to three times. Have you examined the activity of Cohn fraction IV?

Holley: We haven't combined fraction IV with the purified β-globulin fraction, but we have assayed fraction IV directly and in combination with crude fraction III, without finding any interaction.

Clarke: Can you be certain that the factor which stimulates increased growth of SV3T3 cells is not one needed for survival of cells, which are able to pass through the cell cycle without the addition of serum?

Holley: The SV3T3 factor does not give survival of SV3T3 cells by itself. If one adds the SV3T3 factor without serum and then adds fractions from an electrophoretic fractionation of serum, there are fractions from the electrophoresis that give survival. There is something required for survival and also something else that stimulates growth.

Rubin: You said that rat serum is two and a half times more efficient than calf serum, and in your original paper (Holley and Kiernan 1968) you indicated that mouse serum was even more effective. Have you any reason to believe that it is the same material, or even the same *kind* of material, in all these sera that is the active fraction?

Holley: We haven't any evidence either way. Any of them will suffice to grow the cells.

Rubin: How much more efficient is mouse serum? With chicken cells we find that heterologous serum generally becomes toxic in higher concentrations, and for general purposes it is almost essential to use homologous serum.

Holley: Mouse serum is another three times as active as rat serum, with 3T3 cells. We don't know why; my assumption has always been that there is some species-specificity in the structures of the active compounds, so that slightly different compounds are fulfilling the same function. To stimulate a mouse cell you need more of the calf factor than of the equivalent compound from the mouse.

Rubin: What I had in mind was that perhaps many things in homologous serum work that don't work at all in a heterologous serum; that in addition to β-globulin which you find, there might for example be an α-lipoprotein which is active (Rejnek *et al.* 1965). In addition, peptides from enzymic digests (De Luca, Habeeb and Tritsch 1966) and even dextran (Healy and Parker 1966) have been found to be active in replacing sera in supporting cell growth.

Dulbecco: Have you seen any species difference with any purified material, or just with the whole serum?

Holley: If you chromatograph calf serum at pH 3 on a Sephadex G100 column you find activity in the same place as with rat serum, but less activity.

Rubin: Is the bovine β-globulin fraction a lipoprotein fraction, and can some of the peculiar properties like adsorption on charcoal and its great instability at neutral pH be explained by its being a lipoprotein?

Holley: We have no evidence that it is a lipoprotein. The activity is not extracted with ether. Sometimes one can get toxic material back from ether. Jim Bartholomew in my laboratory has prepared lipoproteins from serum and has not found activity in the lipoprotein fraction. All sorts of things stick on charcoal, of course.

Pitts: Has anyone looked for growth factors in ascitic fluid from tumour-bearing mice?

Burger: We examined the effect of ascitic fluid from mice injected with L1210 cells upon confluent 3T3 cells (K. Jacobus and M. M. Burger, unpublished observation). A definite stimulation of growth of confluent 3T3 cells could be observed. However we found a lot of proteolytic enzymes in the ascitic fluid, and this might explain such an effect. Have you tested for enzymic activity in these fractions? Particularly esterase and proteolytic activity would be good candidates, since such activities are quite often resistant to pH 2 or even 100°C. Trypsin, for example, is also resistant to low pH, and so of course is pepsin.

Holley: We haven't tested for enzymic activity yet, but this is something we want to do.

Montagnier: Have you tried denaturing agents like urea on the activity of your factors?

Holley: No, we haven't.

Bergel: If, as you think, your active fraction is a functional protein, could you not utilize the technique now applied to enzymes, namely to link it covalently with carriers such as porous glass (Weetall 1969) or organic macromolecules (see Kay *et al.* 1968) and thus "insolubilize" it? Although the stability may not be increased significantly, such material could be handled with greater ease, especially during assays.

Holley: Eventually we hope to know, and we would like to bind our fraction to some solid support and see if it is active that way. My guess is that it will be.

Pitts: Have you looked at the stability of the active fraction under conditions other than low pH, such as high pH or high salt concentrations?

Holley: We have done quite a bit with whole serum. The activity is stable at moderate temperatures from pH 2 to 11.

Montagnier: Can you make an estimate of the molecular weight of your factor?

Holley: The position on the Sephadex column and the fact that it is non-dialysable suggest that the molecular weight is 10–30 000, but it could be a molecule of higher molecular weight that is adsorbed and delayed to the chromatographic position corresponding to 20 000.

Mitchison: Is the depletion of the serum activity by exposure to cells anything to do with cell activity, or is it purely passive? If you incubate serum with a lot of cells in the cold, for example, will that remove the activity?

Holley: We haven't done this in the cold. It is something that at 37°C requires time; it appears to go on in approximately the same way whether the cells are growing or not, but it always seems to stop. That is, the activity doesn't diminish to zero; it goes to approximately that concentration which is in equilibrium with the density of cells. However, it is very hard to obtain really quantitative data here because the concentration of the activity is low.

Stoker: Have you ever analysed what is left?

Holley: No, because the concentration is too low to assay the fractions after chromatography. We would have to recover the high molecular weight material and then try to fractionate it. Even starting with calf serum alone we have difficulty assaying after the fractionation.

Stoker: So you don't know if the residual activity is due to the same molecule.

Holley: No.

Dulbecco: Is the depletion of the medium the result of an equilibrium between attachment and elution of serum factors? Can you take the cells, remove the medium, put them in a medium without serum and find activity appearing in the new medium?

Holley: Not in our experience with 3T3 cells. The strongest evidence is from wounding cultures. We know that not more than 0·1 per cent serum is needed to get 3T3 cells to crawl out from the edge of a wound; yet if you change to medium without serum and then wound the culture, there is no significant crawling out over 20 hours. From the rate of depletion normally, and the way the depletion curve levels off, if depletion is of a single growth factor that is also produced by the cells, the growth factor would have to be produced at a rate such that 0·1 per cent serum activity would appear in a few hours. However, we know we have a complex situation with several factors. It may be that a trace of one factor is required for the cells to make another.

Rubin: If an equilibrium is to be established, there must be dissociation of the serum molecules which combine with the cells. On this point, we found (F. Steck and H. Rubin, unpublished) that chick embryo cells grown first in calf serum and then in chicken serum plus bovine albumin, continued to release non-albumin bovine antigens for three successive days. Three precipitation bands were found in Ouchterlony gels the first day after the change, and one or two on subsequent days.

Pontecorvo: Dr Holley, what does final cell density really mean? Is it the number of cells per unit of surface? Are the cells smaller or larger?

Holley: 3T3 cells as they get more dense still form a monolayer up to several times the minimum confluent density; you can increase to five times the density that would cover the plate and still have a monolayer. At higher densities the cells pile up.

Pontecorvo: Are the cells of the same volume in both cases?

Holley: As far as we are aware they are the same volume, but they spread more at low cell density.

Rubin: Chicken cells decrease in size when they are more crowded. They can decrease in volume by as much as two times.

REFERENCES

DE LUCA, C., HABEEB, A. and TRITSCH, G. (1966) *Expl Cell Res.* **43**, 98–106.

HEALY, G. M. and PARKER, R. C. (1966) *J. Cell Biol.* **30**, 539–553.

HOLLEY, R. W. and KIERNAN, J. A. (1968) *Proc. natn. Acad. Sci. U.S.A.* **60**, 300–304.

KAY, G., LILLY, M. D., SHARP, A. K. and WILSON, R. J. H. (1968) *Nature, Lond.* **217**, 641.

REJNEK, J., BEDNAŘÍK, T., ŘEŘÁBKOVÁ, E. and PEŠKOVÁ, D. (1965) *Expl Cell Res.* **37**, 65–78.

SHORT, J., ZEMEL, R., KANTA, J. and LIEBERMAN, I. (1969) *Nature, Lond.* **223**, 956.

WEETALL, H. H. (1969) *Nature, Lond.* **223**, 961.

CONDITIONS AFFECTING THE RESPONSE OF CULTURED CELLS TO SERUM

G. D. CLARKE AND M. G. P. STOKER

Department of Transformation Studies, Imperial Cancer Research Fund Laboratories, Lincoln's Inn Fields, London

IT has been known for many years that a relationship exists in layer cultures of normal fibroblasts between local cell density and mitotic index (Willmer 1933), but only recently has there been any study of the mechanism involved. The rates of synthesis of DNA, protein and RNA are now known to be reduced in dense stationary phase cultures compared with those in the logarithmic phase of growth (Levine *et al.* 1965). The saturation density at which growth ceases is characteristic of the cell type concerned (Stoker 1967; Elsdale 1969) but depends on the concentration of serum (Todaro, Lazar and Green 1965; Temin 1966; Holley and Kiernan 1968), the addition of which will cause even static cultures to grow (Todaro, Lazar and Green 1965). Renewed growth may also be stimulated in such cells by subculturing to a lower cell density or by wounding the layer (Todaro, Lazar and Green 1965; Dulbecco and Stoker 1970). This reactivation is not dependent on any injury to the cells, since variation in response has been observed at different densities of cells in the same culture dish (Levine *et al.* 1965), and even in the centre of large colonies compared with the periphery (Fisher and Yeh 1967).

Growth-promoting activity has been demonstrated in the serum of all animals tested and shows little species specificity. It is non-dialysable and may be precipitated with the γ-globulin fraction (Jainchill and Todaro 1970). Although some activity has been found in impure preparations of insulin (Temin 1967) and pituitary hormones (Holley and Kiernan 1968), the effect has not been accounted for quantitatively by that of any known hormone or serum fraction.

The growth of normal fibroblasts also depends on anchorage to a substrate and does not occur readily when the cells are suspended in agar (Macpherson and Montagnier 1964) or methyl cellulose (Stoker *et al.* 1968). The non-growing resting cells in both dense layer and suspension cultures remain in the G1 phase of the cell cycle (Nilausen and Green 1965; Stoker

17

et al. 1968), release from these conditions being followed, after a short lag, by DNA synthesis and mitosis.

Thus, growth in culture is affected by a stimulating activity in serum and by cell density and anchorage. Fibroblasts which have been transformed by tumour viruses are affected less by these factors (Temin and Rubin 1958; Vogt and Dulbecco 1960; Stoker and Macpherson 1961; Todaro, Green and Goldberg 1964; Macpherson and Montagnier 1964).

These conclusions have been based on studies of cells from a variety of animal species, the 3T3 mouse line and BHK21 hamster line being frequently used. The present paper will describe work in our laboratories largely confined to the BHK21 line and its polyoma virus-transformed derivatives. We will first summarize recent observations on the effect of serum under various conditions of culture (Clarke *et al.* 1970) and then describe work on the depletion of serum activity by cell growth, the influence of substances secreted by normal fibroblasts themselves and the inhibitory action of polyanions. In what follows, "cells" are BHK21 clone 13 and "transformed cells" are their polyoma virus-transformed derivatives.

CHARACTERIZATION OF BHK21 CELLS

BHK21 cells grow in layer culture in Eagle's medium (Dulbecco's modification), with 10 per cent serum, to a saturation density of about 10^6 cells per cm^2, compared with 5×10^4 cells per cm^2 for 3T3 cells. In confluent cultures cells are spindle shaped and oriented with respect to one another. We obtain resting cells in the G1 phase of the cell cycle by incubating non-confluent layer cultures of 3×10^5 cells per 50 mm dish for two to three days in 0·5 per cent calf serum medium (Bürk 1967). When this is replaced by serum-free medium containing [^{3}H]thymidine for 20 hours, only a small proportion (approximately 1 per cent) of the nuclei are found to be labelled after radioautography (Clarke *et al.* 1970). The addition of serum increases the number of labelled nuclei roughly proportionately, up to 50 per cent being labelled with 3 per cent serum and 80–90 per cent with 10 per cent serum. All evidence suggests that this procedure measures the first cycle of DNA synthesis, but since the first labelled nuclei appear about 9 hours after the addition of serum, it is an assay of the initiation of the cycle in G1.

When this assay of serum activity was done in cell layers of different density it was found that the slope of the dose–response curve was affected by cell density. The response to low doses of serum was maximal with sub-confluent layers at 3–5×10^5 cells per 50 mm dish and was reduced at higher densities. This manifestation of density-dependent inhibition was not observed with 10 per cent serum. The response was also reduced at lower cell densities, probably through lack of conditioning activity (see below).

We also investigated the response to serum of cells suspended in medium containing 1 per cent Methocel (Dow Chemical Corporation, "4000 centipoises") and found that they were about 60 times less responsive than in layer cultures. However, 30 per cent of the cells formed colonies in suspension when the serum concentration was raised to 50 per cent. Although the failure to grow in agar appears to be a reflection of the decreased sensitivity to serum in suspension, there are other factors involved, which will be discussed later in relation to the effects of polyanions.

The basic problem to be resolved concerning the stationary phase of normal cells in layer culture is the relative importance, for any particular cell type, of the exhaustion of growth-stimulating substances from the serum and the so-called "contact inhibition of growth". Published findings appear at first sight to be in conflict, and we have attempted to obtain further information with our cell system.

DEPLETION OF SERUM ACTIVITY BY GROWTH

Medium which has been incubated on stationary phase cultures of 3T3 cells is reported to support the growth of a second culture to full confluency (Todaro, Lazar and Green 1965) but medium in which a culture has grown to confluency only supports cell growth to a much lower saturation density (Holley and Kiernan 1968). Chicken fibroblasts in secondary culture also appear to remove growth activity from the medium (Temin 1969).

In order to avoid any artifacts due to decreased concentration of small molecular components of the medium during similar depletion experiments, we used a method suggested by Ted Gurney (personal communication 1970). Cells were grown in 10 per cent serum medium for 3–4 days and the medium was then removed and dialysed exhaustively against fresh 10 per cent serum medium. Fresh cells of the same type were seeded into the used medium for a further growth period, and this procedure was repeated until the cells failed to increase in numbers significantly in the growth period (Table I). The medium was then dialysed again and assayed on resting non-confluent cultures by the standard procedure. Medium subjected to a similar regimen, but without cells, showed no significant loss of activity, while the depleted medium had a residual activity equivalent to that of 1–2 per cent serum. Since this residual activity was lost on dilution

TABLE I

DEPLETION OF GROWTH ACTIVITY IN CULTURE

BHK21 and Py* cells seeded at $1 \cdot 2 \times 10^7$ per roller bottle in 200 ml of medium containing 10 per cent serum. Medium removed and cells counted after 3–4 days at 37°C. Medium treated as in text. 3T3 cells seeded at 4×10^6 per bottle.

Expt no.	Cell type	Cell numbers $\times 10^{-7}$ in consecutive cultures				
		1	2	3	4	5
1	BHK21	8·4	2·1	—	2·0	—
2	BHK21	17·5	0·9	0·7	—	—
3	BHK21	15·4	1·46	0·67	—	—
4	BHK21	16·6	1·64	0·73	—	—
5	Py*	10·5	2·35	3·67	2·0	1·58
6	3T3	2·0	2·0	1·05	—	—

* BHK21 cells transformed with polyoma virus.

in the same way as serum itself, there was no evidence of accumulation of a non–dialysable inhibitor (Fig. 1). Very similar results were obtained when medium was depleted by 3T3 cells instead of BHK21 cells. This would support the thesis that actively growing cells exhaust growth activity, as shown by Holley and Kiernan (1968). The activity of medium depleted by transformed cells did not decrease as expected on dilution (Fig. 1), but the significance of this anomaly is not yet clear.

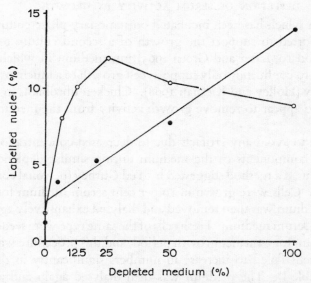

FIG. 1. Depleted media from normal (–●–●–) and transformed (–o–o–) BHK21 cultures (Table I, 2 and 5 respectively) were diluted with serum-free medium and titrated on resting normal cells by radioautography after incubation for 20 hours at 37°C with [³H]thymidine (see text). Over 300 cells each counted on duplicate dishes; each curve is mean of two separate titrations.

The residual activity persisted after continued exposure to cells but it cannot be assumed that this is due to a remaining 10–20 per cent of the activity in the original serum, because the cells may also release conditioning factors which affect the assay.

ACTIVITY RELEASED BY BHK21 CELLS

Our attention was originally drawn to conditioning factors by the very high sensitivity to serum of cells emigrating from confluent sheets into wounds made in the culture. The proportion of these cells which were stimulated by a low dose of serum was higher than would be expected from the reduced cell density alone, and it was then shown that this high sensitivity to serum also affected superadded non-confluent cells in the same culture at some distance from the edge of a confluent sheet seeded on half of the dish (Figs. 2a and b), without any evidence of a gradient of response approaching the edge. This suggested that the confluent cells were releasing an activity into the medium which enhanced the response of non-confluent, but not that of confluent, cells to serum. Similar results are obtained whether or not the cell sheet is wounded (Fig. 2c), and such an activity was then demonstrated in serum-free medium alone after exposure to confluent cell layers.

In addition to having an enhancing, or serum-sensitizing, activity which alters the slope of the dose–response curve, conditioned serum-free medium, particularly after exposure at a high cell-to-medium ratio, will stimulate non-confluent cells to incorporate thymidine even in the absence of any added serum. This serum-replacing activity may also be demonstrated by stimulation of incorporation in wound cells in small volumes of serum-free medium or in medium containing a low concentration of gamma-globulin-depleted serum (Figs. 2d and e). These two types of conditioning activity, serum sensitizing and serum replacing, can be assayed separately (Fig. 3), but we do not yet know whether two separate factors are responsible.

The activities are related to the number of conditioning cells and even after exhaustion of the medium and repeated washing of cell layers the combined activities are still continuously released by cells for at least six days. The serum-replacing activity might still come from serum factors stored since the earlier propagation of the cells in serum, but the sensitizing activity is presumably contributed by the cells themselves. So far, no progress has been made in purifying the factors responsible, but the activities are due, at least in part, to non-dialysable material, and they are inactivated in 30 minutes at $56°C$ (pH $7 \cdot 4$–$7 \cdot 6$).

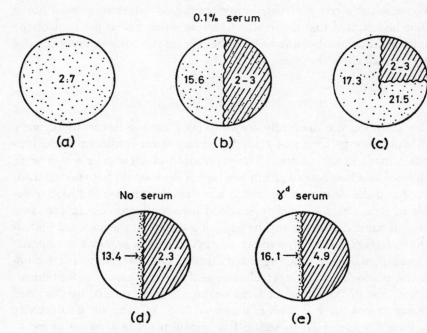

FIG. 2. (a), (b) and (c). Stimulation of thymidine incorporation by non-confluent BHK21 cells (dots) in the presence of a confluent cell layer (hatched). 10^6 cells were allowed to attach to half the surface of a 50 mm dish and left for 3 days in 2 per cent serum medium, washed three times in serum-free medium and left in serum-free medium for 16 hours. 5×10^5 cells from serum-depleted cultures were then added in 5 ml of fresh 0·1 per cent serum medium to the confluent half-cultures (b); to similar cultures in which a further half of the confluent sheet was wiped off (c); and to bare dishes (a). The cultures were incubated for 20 hours with [^{3}H]thymidine before radioautography. Figures are percentages of labelled nuclei, based on counts of at least 400 cells in replicate cultures.

(d) and (e). Thymidine incorporation in emigrating wound cells in small volumes of serum-free medium or gamma-depleted-serum medium. Confluent cultures in 50 mm dishes left in 0·5 per cent serum for three days, washed once in serum-free medium and incubated for 24 hours in 1 ml volumes of fresh serum-free medium or 0·8 per cent gamma-depleted-serum medium. A strip of confluent cell layer was then wiped off and [^{3}H]thymidine added without medium change for a further 20 hours before radioautography. Figures give percentages of labelled nuclei based on counts of 200 cells in each of five replicate cultures.

It is obviously important to determine whether the BHK21 cells themselves synthesize these growth factors, and if so to characterize them. Until this is possible, the main interest lies in the fact that the confluent, activity-releasing cultures are insensitive to their own products, and only respond when released by reduction in density; for example, by wounding. Since the growth-stimulating materials come from the cells, lack of response cannot be due to a simple failure in uptake. It is additional evidence for some other over-riding but unknown role of contact in growth regulation.

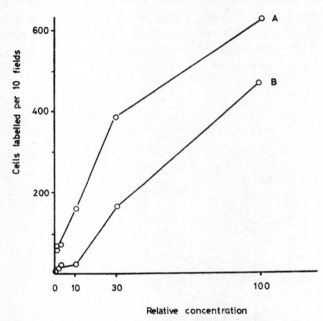

FIG. 3. Activity in concentrated conditioned medium which stimulates incorporation of [³H]thymidine by non-confluent cells in the absence of added serum. Medium without serum conditioned by exposure for one day to confluent layers in roller bottles (30 ml per 10⁸ cells). Conditioned medium diluted with fresh medium as shown on the abscissa is then added, (A) with 0·5 per cent serum or (B) without serum to serum-depleted non-confluent cultures (3 × 10⁵ cells per dish after 3 days in 0·5 per cent serum medium). [³H]thymidine added for 20 hours before radioautography. Labelled cells counted for 10 fields (approx. 1200 cells); mean of two replicate cultures shown.

COOPERATIVE ACTION OF INSULIN

Apart from conditioned medium, enhancement of the response to low doses of serum is also obtained with insulin. Chicken fibroblasts are known to respond to insulin (Temin 1967) but the response to serum cannot be accounted for quantitatively by its insulin content. We found that insulin was active in stimulating uptake of thymidine in resting cells, provided that the serum concentration was greater than 0·25 per cent (Fig. 4). This suggests that the hormone also acts cooperatively; for example, it might increase uptake of materials in serum. It would seem advisable, therefore, that assays of any purified fraction from serum should be made both with and without added insulin.

Although diffusible substances can enhance the response of cells at low density to serum, unresponsiveness in confluent cultures has not been altered other than by reducing the density. This unresponsiveness might be due to a mechanism involving actual contact between plasma membranes, but alternatively could involve membrane-associated extracellular macromolecular material secreted by the cell.

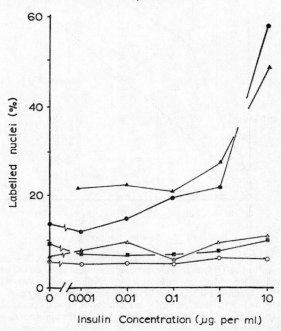

FIG. 4. Effect on [³H]thymidine uptake of medium containing no serum (–o–o–) or 0·125 per cent (–■–■–), 0·25 per cent (–△–△–), 0·5 per cent, (–●–●–) and 1 per cent (–▲–▲–) serum. Insulin (Allen and Hanbury, 23·5 units per mg) added as shown; medium then titrated on resting normal cells by radioautography after incubation for 20 hours at 37°C with [³H]thymidine. Points are average of counts on over 300 cells on each of duplicate dishes.

EFFECTS OF POLYANIONS

Since sulphated mucopolysaccharides, together with collagen, are the main products secreted specifically by connective tissue cells, the known effects of polyanions on normal fibroblasts in culture will now be briefly considered. Temin discovered the inhibitory effect of a sulphated polysaccharide, dextran sulphate, while investigating the action of an agar overlay on the growth of layer cultures of chicken fibroblasts (Temin 1966). His results were compatible with the conclusion that polyanions combined with the growth-stimulating activity in serum, thereby rendering it unavailable to the cells.

We find that both dextran sulphate and heparin, the sulphated polysaccharide secreted by mast cells, inhibit the effect of serum on resting cells (Table II). The effect is not a simple toxicity since the cells do not enter

TABLE II

EFFECT OF POLYANIONS ON RESTING CELLS

Polyanions are added to medium containing serum before its addition to resting cells at 3×10^5 per dish for titration by radioautography (see text). Figures are percentages of labelled nuclei in 600 counted in duplicate dishes.

		Polyanions at 50 µg per ml			
		Dextran sulphate			
Percentage serum	No polyanion added	M.W.: 5×10^5	2×10^6	5–40×10^6	Heparin
2·0	39·4	7·3	5·0	17·9	13·6
5·0	85·0	25·0	22·4	39·1	41·5

DNA synthesis or multiply, but do not die, their number remaining constant for many days. The dose–response relationship is similar to that reported for chicken fibroblasts (Temin 1966), low concentrations of polyanions causing a marked inhibition, but further increments having a decreasing effect (Fig. 5). Pre-incubation of dextran sulphate with medium containing serum does not increase the inhibition but pre-incubation of resting cells with the polyanion decreases the sensitivity of cells to serum (G. D. Clarke and A. Ludlow, unpublished). When cells are pre-incubated with serum for 3 or 6 hours before the addition of dextran sulphate, the inhibitory effect is significantly diminished even though the incorporation of [³H]thymidine into the cells is not appreciable until 9 hours after adding serum. Clearly, by 3 hours some cells have become resistant to the effect of dextran sulphate, which does not then prevent their entry into DNA synthesis. These results suggest that the polyanion may react with the cells rather than with any substance in the serum; kinetic experiments using labelled dextran sulphate might resolve this.

The presence of substances with an activity comparable to that of dextran sulphate clearly may inhibit the growth of normal BHK21 cells in agar. It has been reported that these cells will grow in agarose, which contains far less acidic polysaccharide than does agar, and that their growth is inhibited by the addition of low concentrations of dextran sulphate (Montagnier 1968). Their plating efficiency in agar is increased by conditioned medium (Tjötta, Flikke and Lahelle 1967) and by insulin (I. A. Macpherson, personal communication). Nevertheless, spreading on a surface either increases the sensitivity of cells to serum, or decreases the effect of inhibitors. C. O'Neill (personal communication) has shown that while rounded, unattached, but viable BHK21 cells in suspension fail to multiply, similar cells a few micrometres away, which have attached to glass fibres or beads, grow to form colonies.

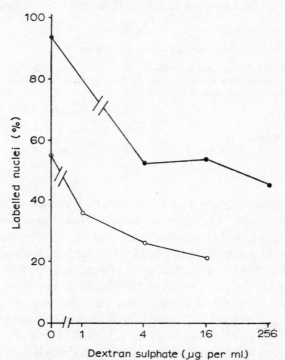

FIG. 5. Effect of medium containing 2 per cent (-o-o-) and 8 per cent (-●-●-) serum and dextran sulphate (M.W. 2×10^6) as indicated, titrated on resting normal cells by radioautography after 20 hours incubation with [³H]thymidine (see text).

Transformed cells will grow in suspension in agar and are insensitive to inhibition by polyanions (Montagnier 1968). This behaviour may reflect the reduced requirement for serum reported for transformed cells in layer culture, but could also be related to their changed surface structure.

POLYOMA-TRANSFORMED BHK21 CELLS

Experiments with both SV40-transformed and normal 3T3 cells (Holley and Kiernan 1968) and Rous sarcoma virus-transformed chicken and rat cells (Temin 1969, 1970) certainly suggest that these transformed cells utilize serum more efficiently than normal cells. When transformed BHK21 cells are transferred to serum-free medium, about 50 per cent of the cells continue to incorporate thymidine, whether they are suspended or in layer culture. The proportion labelled is hardly affected if serum is again added in varying concentrations. This incorporation of thymidine in the absence of added serum cannot be studied for more than 24 hours because the number of viable cells falls. Thus the cells cannot be depleted of serum and the possibility remains that the incorporation could be due to a carrying over of serum activity from the previous culture. However, transformed BHK21 cells in suspension cultures containing 10 per cent serum incorporate thymidine freely, and divide, whereas normal cells do not. Moreover, when serum-depleted normal BHK21 cells in serum-free medium are abortively transformed by exposure to polyoma virus, thymidine incorporation is initiated in most of the cells in the absence of added serum (J. Taylor-Papadimitriou, M. Stoker and P. Riddle, unpublished). This indicates that transformation removes most or all of the need for serum for the initiation of the cell cycle in resting cells. It should be remembered, however, that other factors in serum may be required at later stages in the cycle (Dulbecco 1970), and we have no evidence that transformation removes the need for these.

SUMMARY

The failure of hamster fibroblasts of the BHK21 cell line to multiply in suspension or dense monolayer cultures has been analysed in terms of the interaction between serum and cells required for the initiation of DNA synthesis. Interactions between cells, or between cells and substrate, affect the response to serum. Thus saturation density, for example, is determined not only by the supply and removal of serum but also by the changing requirement of individual cells as the density changes.

Proliferation of normal cells in 10 per cent serum medium reduces the ability of the serum to support growth. The residual activity in exhausted serum, when diluted and titrated on resting cells, behaves as if it is due to 1–2 per cent serum, and there is no evidence of an inhibitor being produced by the cells.

The interaction is complicated by conditioning substances released by the cells themselves, which are of at least two types, causing (1) a serum-like activity which initiates DNA synthesis in the absence of added serum and (2) a serum-sensitizing activity which enhances the response of cells to added serum. Polyanions prevent normal cells from reacting to serum and thus from entering S phase; they do not inhibit DNA synthesis or kill the cells.

BHK21 cells transformed by polyoma virus do not require added serum for DNA synthesis, but they cannot be propagated in the absence of serum. Presumably transformed cells, and normal cells, require other factors in serum for completing the growth cycle.

Acknowledgements

The valuable assistance of Miss M. Thornton and Miss A. Ludlow is gratefully acknowledged.

REFERENCES

BÜRK, R. R. (1967) In *Growth Regulating Substances for Animal Cells in Culture*, pp. 39–50, ed. Defendi, V. and Stoker, M. Philadelphia: Wistar Institute Press.

CLARKE, G. D., STOKER, M. G. P., LUDLOW, A. and THORNTON, M. (1970) *Nature, Lond.* **227**, 798–801.

DULBECCO, R. (1970) *Nature, Lond.* **227**, 802–806.

DULBECCO, R. and STOKER, M. G. P. (1970) *Proc. natn. Acad. Sci. U.S.A.* **66**, 204–210.

ELSDALE, T. (1969) *Ciba Fdn Symp. Homeostatic Regulators*, p. 272. London: Churchill.

FISHER, H. W. and YEH, J. (1967) *Science* **155**, 581–582.

HOLLEY, R. W. and KIERNAN, J. A. (1968) *Proc. natn. Acad. Sci. U.S.A.* **60**, 300–304.

JAINCHILL, J. L. and TODARO, G. J. (1970) *Expl Cell Res.* **59**, 137–146.

LEVINE, E. M., BECKER, Y., BOONE, C. W. and EAGLE, H. (1965) *Proc. natn. Acad. Sci. U.S.A.* **53**, 350–356.

MACPHERSON, I. and MONTAGNIER, L. (1964) *Virology* **23**, 291–294.

MONTAGNIER, L. (1968) *C. r. hebd. Séanc. Acad. Sci., Paris, Sér. D.* **267**, 921–925.

NILAUSEN, K. and GREEN, H. (1965) *Expl Cell Res.* **40**, 166–168.

STOKER, M. G. P. (1967) *Curr. Top. Dev. Biol.* **2**, 107–128.

STOKER, M. and MACPHERSON, I. (1961) *Virology* **14**, 359–370.

STOKER, M., O'NEILL, C., BERRYMAN, S. and WAXMAN, V. (1968) *Int. J. Cancer* **3**, 683–693.

TEMIN, H. (1966) *J. natn. Cancer Inst.* **37**, 167–175.

TEMIN, H. (1967) *Int. J. Cancer* **3**, 771–787.

TEMIN, H. (1969) *J. cell. Physiol.* **74**, 9–16.

TEMIN, H. (1970) *J. cell. Physiol.* **75**, 107–120.

TEMIN, H. and RUBIN, H. (1958) *Virology* **6**, 669–688.

TJÖTTA, E., FLIKKE, M. and LAHELLE, O. (1967) *Arch. ges. Virusforsch.* **23**, 288–291.

TODARO, G. J., GREEN, H. and GOLDBERG, B. (1964) *Proc. natn. Acad. Sci. U.S.A.* **51**, 66–73.

TODARO, G. J., LAZAR, G. K. and GREEN, H. (1965) *J. cell. comp. Physiol.* **66**, 325–333.

VOGT, M. and DULBECCO, R. (1960) *Proc. natn. Acad. Sci. U.S.A.* **46**, 365–370.

WILLMER, E. N. (1933) *J. exp. Biol.* **10**, 323–339.

DISCUSSION

Macpherson: I should like to add some comments on the effect of insulin in agar suspension cultures. The addition of 1–10 µg of insulin to Difco-Bacto agar medium enables BHK21 and Nil2 hamster cells to form colonies. Growth slows down after two or three days but can be re-stimulated by adding more insulin in liquid medium to the agar top layer. BHK21 cells transformed by polyoma or Rous sarcoma virus, which grow in agar, are not further stimulated to grow faster or to plate with a higher colony-forming efficiency by the addition of insulin.

The insulin effect does not seem to be due to its contamination with zinc or glucagon since these substances over a wide range of concentrations do not induce the growth of BHK21 in agar.

Experiments by L. Brennerman and myself (unpublished data) on the uptake and incorporation of labelled glucose into BHK21 and transformed BHK21 suggested that insulin did not enable BHK21 cells to grow in suspension by stimulating their utilization of glucose.

Bergel: In addition to dextran derivatives, have you tested polyphosphates or other polyelectrolytes? Sela and Katchalski (1959) have studied the inhibitory effects of polyamino acids on a variety of cells and viruses, and Regelson (1968) those of polyanions. There appear to be differences in activity between one and the other type of compound.

Clarke: Polycations that we have tested, unlike polyanions, have been toxic to resting BHK21 cells in layer culture. There is some evidence, both in the literature and with our system, that they are less toxic to cells in suspension.

Rubin: An overlay of sulphated agar is highly inhibitory to cell growth whereas agarose, which has a much reduced sulphate content, is not inhibitory.

Montagnier: We have tried several groups of acid polysaccharides on BHK21 cells growing in agarose. Only sulphated polysaccharides (such as dextran sulphate and heparin) were found to be inhibitory. Hyaluronic acid of various molecular weights, even at high concentrations, had no such effect; instead, it caused some stimulation of growth.

Clarke: Unlike chicken cells (Temin 1966), resting BHK21 cells in layer culture do not seem to be inhibited by an agar overlay.

Rubin: We see a gross effect of unpurified agar on the growth of chicken cells, whereas agarose has no effect.

Mitchison: I am reminded of the action of dextran sulphate and heparin, which cause recirculating lymphocytes to lose their way. A particularly effective agent for doing this is polymethacrylic acid (Ormai and De Clercq 1969). It would be interesting to see if there is a parallel between the effects of these agents in this *in vitro* system and on lymphocytes *in vivo*.

Clarke: We have not tested this. I should have mentioned that polyanions have been reported also to cause a marked stimulation of pinocytosis by mouse macrophages (Cohn and Parks 1967).

Rubin: Dr Clarke mentioned Dr Ted Gurney's experiments on depletion of the macromolecular serum factors in the medium. Gurney found (unpublished results) after six successive rounds with large numbers of

2*

normal chicken cells ($1 \times 10^6 - 4 \times 10^6$), the medium being dialysed after each growth period, that there was no evidence of loss of activity in the medium. However, with Rous sarcoma cells, running the medium in polyacrylamide gel to detect the loss of specific proteins in the serum, he occasionally found total loss of two minor bands, but he could not reproduce this. This brings up the question of whether some of the depletion, rather than being due to adsorption of something from the medium, is due to destruction of some components of the medium.

Clarke: We have no evidence yet to distinguish between depletion and breakdown by BHK21 cells.

Rubin: The obvious possibility is that the cells may make proteases.

Dulbecco: You mentioned the possibility of inhibitors being responsible for the crowding effect, but the wounding experiment surely shows that this is extremely unlikely?

Clarke: The total cell volume is very low compared with that of the culture medium, so that molecules are likely to be diluted out rapidly after secretion. One would not, therefore, expect to detect inhibitor activity easily in used medium, and the difficulty would be greater if the inhibitor had an attraction towards the cell surface.

Dulbecco: Michael Stoker and I once tried to do an experiment to answer this particular question. We thought of the possibility that a gradient of some substance produced or eliminated by the cells may be formed in the medium in contact with the cells in a wounded cell layer. We decided to see what the effect would be of destroying the gradient by a very strong convection current. We just looked at cell movement in 3T3 cells. There was no effect of convection currents on the cell movements, which occurred quite normally. So the idea of factors which are released in the medium and are short-lived does not explain the characteristics of movement in cell layers (such as the orientation away from the cell layer after wounding), because the convection velocity was probably a thousand times greater than the average diffusion rate even for small molecules.

Clarke: It might then be necessary to postulate that any theoretical inhibitor had some affinity for the cell surface.

Rubin: I don't think you have to worry too much about desorption of serum, because in our experiments (F. Steck and H. Rubin, unpublished) the amount desorbed on each successive day was less and in one experiment by the fourth day there was none; it was not constant by any means. Furthermore, we have shown that cells synthesize and release a variety of proteins which give as complex a picture upon gel electrophoresis as does serum (Halpern and Rubin 1970). Some components require immunoelectrophoresis to distinguish them from gamma globulin and albumin, although they are not in fact gamma globulin and albumin. These proteins were shown to be also synthesized *in vivo*, and are not turned on in response to cultivation *in vitro*. Finally, at least to some extent part of this sounds like the old conditioned medium story (Rubin 1966); it's clear that cells make a factor in the presence of serum which enhances the growth of small numbers of cells.

Taylor-Papadimitriou: Does the factor that you obtain just stimulate DNA synthesis, or is it also active in stimulating mitosis, and does it help survival?

Stoker: I don't know; so far we have only looked at initiation of thymidine incorporation.

Dulbecco: What is the state of the cell layer in these conditions? It could be that the material is a disintegration product of the cells. There's the additional technical point that these conditions would not be adequate to show whether the material was sedimentable, because there would be a very small quantity and unless centrifugation was done in a density gradient the sedimented material would be stirred up by convection when the centrifuge stops.

Stoker: In the absence of serum the cell sheet does not look too bad for one or two days, but you can't keep the BHK21 cell sheets for longer than that; the sheets start to retract. The experiments on the length of time the release goes on were done in gamma-depleted serum, the equivalent of 0·5 per cent serum. These cell sheets look quite good for at least a week. This is why we emphasized that we were assaying combined activity, because there may still be activity left in the gamma-depleted serum, and we may then be mainly looking at the interaction factor, but they could still be cell breakdown products; even with a completely healthy sheet one could not rule this out.

Rubin: You could test cell extracts for the activity.

Stoker: Yes. We have stuck to medium so far, because we want to get as low a protein concentration as possible.

Taylor-Papadimitriou: I did once test extracts of mouse embryo fibroblasts prepared by scraping them off the glass and freezing and thawing. The extract, at a final dilution of 1 to 10, stimulated about 30 per cent of the cells to synthesize DNA. Of course, since the cells had been grown in serum one cannot be sure whether the effect was due to serum released from the cells.

Montagnier: Fibroblasts release collagen into the medium; therefore it is present in conditioned medium. Have you used collagen to see if you could get the same effect?

Stoker: No.

Warren: Has anyone treated preparations containing growth-controlling factors with fractions of membranes, especially those of the surface of the cell, to see if the factors can be adsorbed from solution?

Mitchison: Isn't that very much what your cold cells would have been doing?

Stoker: We haven't done this. All we know is that the activity does not appear in cultures kept at 4°C.

Pitts: You mentioned the possible heterogeneity of your cells. Have you tried to select for cell variants which respond differently to the factor or factors?

Stoker: Dr Shodell is trying to do this now.

Shodell: I have been trying to get variants of BHK cells that will grow well in low concentrations of serum. However, the sort of heterogeneity indicated here is one in which cells are able to enter the S phase (i.e. incorporate tritiated thymidine) under limiting serum conditions. Whether these cells can then continue, under these limiting serum conditions, to complete mitosis and initiate further rounds of division, as would be necessary for the sort of selection suggested, is probably unlikely.

Clarke: Attempts to select non-requiring variants for serum factors may be complicated by the need for several activities, which have been discussed by Dr Dulbecco (1970). However, dextran sulphate does not, in the presence of 4 per cent serum, inhibit survival and if it only prevents the interaction which facilitates the G1 to S step, then such a medium might be used to select mutants independent of serum for this step.

REFERENCES

COHN, Z. A. and PARKS, E. (1967) *J. exp. Med.* **125**, 213–232.
DULBECCO, R. (1970) *Nature, Lond.* **227**, 802–806.
HALPERN, M. and RUBIN, H. (1970) *Expl Cell Res.* **60**, 86–95.
ORMAI, S. and De CLERCQ, E. (1969) *Science* **163**, 471–472.
REGELSON, W. (1968) *Adv. Chemother.* **3**, 303.
RUBIN, H. (1966) *Expl Cell Res.* **41**, 138–148.
SELA, M. and KATCHALSKI, E. (1959) *Adv. Protein Chem.* **14**, 391.
TEMIN, H. (1966) *J. natn. Cancer Inst.* **37**, 167–175.

FACTORS CONTROLLING THE MULTIPLICATION OF UNTRANSFORMED AND TRANSFORMED BHK21 CELLS UNDER VARIOUS ENVIRONMENTAL CONDITIONS

L. Montagnier

Institut du Radium, Faculté des Sciences, Orsay

THE choice of a permanent cell line such as the BHK21 line of hamster fibroblasts to study the control of cell division *in vitro* has both advantages and disadvantages.

Among the former are the readiness with which one can obtain large quantities of a clonal, hence homogeneous, population of cells; the ability of the line to be transformed by several types of oncogenic viruses, both RNA- and DNA-containing, and therefore the possibility of making comparative biochemical studies on untransformed and virally transformed cells.

The disadvantages are well known: although these cells retain some of the characteristics of "normal" fibroblasts (by "normal", I mean fibroblasts of primary or secondary explantation) they are less sensitive to density dependent inhibition, as shown by their ability to grow in multilayers without change of medium. Moreover, these cells can evolve "spontaneously" and give rise to variants of more transformed tumorigenic cells—for one may consider them to have already undergone a first stage of transformation, which can be defined by accurate criteria as will be later shown, as can the further stages of transformation.

In a search for correlations of biological properties with biochemical changes at the molecular level, the advantages of using a permanent cell line clearly outweigh the disadvantages. Moreover, the latter can be minimized by frequently retesting and eventually recloning the control cells.

Our first interest in this system was to explain why virally transformed BHK21 cells grew when suspended in a nutrient medium gelled with agar, whereas untransformed BHK and normal fibroblasts could not divide

(Montagnier and Macpherson 1964). Subsequent observations (Montagnier 1966) showed that the repression of division was directly related to the spherical shape of the suspended cells and not to a toxic effect of the agar, since when attached to fragments of glass or cotton incorporated fortuitously in the agar, they were able to divide. The phenomenon has been studied in detail by Stoker and associates (1968) and has been termed "anchorage dependence" by them.

So far, this anchorage dependence is absolute for embryonic fibroblasts from primary and secondary explantation, and also for human diploid lines (Montagnier et al. 1967). In contrast, we found that the anchorage dependence of BHK21 cells could be abolished when certain conditions were fulfilled:

(1) The gel must be freed of the sulphated polysaccharides (SPS) contained in the bacteriological agar, which are strongly inhibitory. This can be attained by washing the agar thoroughly, by complexing the SPS with DEAE-dextran, or by using the agarose fraction of agar, which is devoid of SPS.

(2) The cell concentration must be sufficiently high—5×10^4 cells in a 60 mm Petri dish, or more. At lower concentrations (10^4), the cloning efficiency varies with the batch of serum used: with some batches (calf serum or horse serum, at a concentration of 10 per cent) the cloning efficiency is still high (20 to 30 per cent), while with others it drops abruptly. In the latter case, normal cloning efficiency can be restored by adding insulin and collagen (Montagnier 1970a; Sanders and Smith 1970). The growth of the colonies is also more rapid in the presence of these macromolecular factors, since the colonies formed are bigger. Insulin also makes the cells slightly more resistant to SPS: this fact may explain the observation of Macpherson (personal communication) that in the presence of insulin BHK cells can grow in ordinary agar.

At the beginning of these studies, we used the MEM medium of Eagle modified by Macpherson and Stoker (1962) with 10 per cent of Bactotryptose phosphate (BTP) broth. We observed later on that the cloning efficiency was improved by using a less rich medium, the basal medium of Eagle (BME) with a lower concentration of bicarbonate and no BTP. The lower pH obtained with this medium when it is equilibrated with 5 per cent CO_2 may be responsible for this improvement.

In contrast, the small molecule factors of serum are required as well as its macromolecular factors: the cloning efficiency is very low in the presence of dialysed serum and is not fully restored with insulin.

It should be pointed out that the requirements for insulin, collagen and undialysed serum are specific for this type of culture: they do not exist

TABLE I

A SUMMARY OF THE CHARACTERISTICS OF BHK21 CELLS AT DIFFERENT STAGES OF TRANSFORMATION

Clone prototype	Stage	Morphology and orientation	Growth		Growth in agarose + dextran sulphate 10 µg/ml or ordinary agar		Requirement for non-essential amino acids	
			In agarose	Insulin and collagen-stimulated	With AMP	Without AMP	L-Asparagine	Serine
BHK21/13	1a	Long fibroblasts, in parallel	+	+	–	–	+ or –	–
BHK21/13/9	1b	Shorter fibroblasts, less parallel	+	–	–	–	+ or –	±
Ag4–Ag5	2a	Epithelioid	+	–	+	–	+ or –	+
RB12	2a	Epithelioid	+	–	+	–	+ or –	+
RS2–RS3	2a	Short fibroblasts, more or less parallel	+	–	+	–	+ or –	+
Py21, Py39, C13/LPy	2a	Short fibroblasts, random	+	–	–	–	+ or –	+
C13/HSV	2a	Spindle shaped and round cells	+	–	+	–	+ or –	+
BHK21/13/SPy, BHK21/13/9/LPy	2b	Short fibroblasts and round cells, random	+	–	+	+	+ or –	+

for anchored cells growing on a support, at least in the same range of cell concentration. We thus have an example of growth factor requirements being specific for certain environmental conditions.

REQUIREMENTS OF BHK21 CELLS AT FURTHER STAGES OF TRANSFORMATION

A spontaneous evolution occurring in serially passaged cells leads to the emergence of variants with different growth requirements. One can therefore define several stages of transformation, each characterized by its specific growth requirements (Table I). This definition is not only operational, since these stages correspond to a progression of the cells towards malignancy, as indicated by other characteristics of growth in monolayers and tumorigenicity (Montagnier 1970a, b).

If one terms stage 1a the stage at which BHK21 cells are from early passages, three further steps can be defined:

Stage 1b. This stage is frequently reached when confluent cells are passaged serially: BHK cells at this stage become capable of growing in agarose at low concentrations, without insulin and collagen. However, their sensitivity towards sulphated polysaccharides is close to that of cells in 1a. Fibroblasts at this stage are shorter and less well oriented than in 1a.

Stage 2 is defined by the capacity to grow in the presence of the sulphated polysaccharides contained in agar or in agarose containing 10 to 15 μg/ml of dextran sulphate. The easiest way to obtain cells at this stage is to transform them by oncogenic viruses. It is more rarely attained spontaneously. Two subclasses of stage 2 can be distinguished.

In *stage 2a*, growth in the presence of sulphated polysaccharides requires absolutely the addition to the medium of adenine nucleotides (usually provided by BTP). In agarose the number and size of colonies is the same, whether adenine nucleotides are present or not. All the clones tested from cells transformed by RNA oncogenic viruses (RSV, Bryan or Schmidt-Ruppin strains; HSV), or "spontaneously" transformed, are at this stage, as are a fraction of the clones transformed by polyoma virus.

For cells at *stage 2b*, the adenine nucleotides are no longer required for growth in the presence of SPS. Cells transformed by the SP variant of polyoma virus, or cells initially at stage 1b and transformed by all strains of polyoma virus, are at this stage. Cells transformed by RNA viruses and passaged serially also tend to reach this stage (see Fig. 1).

Cells at either stage 2a or 2b are highly tumorigenic (DT50 between 10^2 and 10^4). Their growth in monolayers is randomly oriented, but there are some variations, depending on the transforming agent. Cells at stage 2a show more adhesiveness than polyoma cells at stage *2b*; the latter are generally round and are loosely attached to the glass or plastic support.

In addition, the use of dialysed serum has enabled us to detect a new amino acid requirement for the growth in all conditions of cells at stage *2b*: that for L-serine (Montagnier 1970*a*).

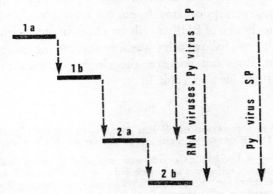

FIG. 1. A diagram of the multi-step transformation of BHK21 cells.

The conclusions to be drawn from these studies are:

(1) Transformation of BHK cells is not an all-or-none phenomenon; one can speak in this case of a multi-sequential transformation, comparable to the progression of transplanted tumours *in vivo*.

(2) There are a number of growth effectors, which differ according to the stage of transformation. For a cell at a given stage of transformation, individual factors are required or not, depending on the environmental conditions. Are these factors needed for other types of cells under the same culture conditions?

(3) Preliminary experiments indicate that adenine nucleotides are also important in other cell systems: spontaneously transformed cells derived from a permanent line of rat fibroblasts have the same behaviour as BHK cells at stage *2a*. Similarly, the growth in agarose of a continuous line of human lymphoblasts, MICH cells (a gift from Dr D. Viza, Searle Laboratory, England), is greatly improved by 5'-AMP.

The next question arises: what is the site of action of these effectors?

THE CELL SURFACE AS THE PRIMARY SITE OF ACTION OF MACROMOLECULAR EFFECTORS

There are some indications that at least some of the effectors described act primarily at the cell surface. This particularly concerns the role of sulphated polysaccharides and that of adenine nucleotides. The experimental evidence can be summarized as follows:

(1) *Role of sulphated polysaccharides*

The inhibitory effect of SPS on cells at stage *1* is maximal when fractions of high molecular weight (dextran sulphate in the range of 5×10^5–2×10^6) are used. Moreover, it does not depend on an irreversible binding to the cell surface: incubation of BHK21 cells for 1 hour in the presence of 10 μg/ml of dextran sulphate, followed by washing and plating in agarose without the polyanion, does not decrease the cloning efficiency of these cells (Montagnier 1968; see also Table II).

TABLE II

EFFECT OF PREINCUBATION OF BHK21 CELLS WITH DEXTRAN SULPHATE AND DEAE-DEXTRAN ON THEIR PLATING EFFICIENCY IN AGAROSE

5×10^5 trypsinized BHK21/13 cells were suspended in 1 ml of BME medium with 10 per cent calf serum and incubated in this medium with or without dextran polymers for 1 hour at 37°C; cells were then centrifuged and plated in agarose medium, at a concentration of 10^5 cells per 60 mm Petri dish.

Treatment	Number of colonies*	Percentage of control
Control	10 800	100
+Dextran sulphate (Mol. Wt. 5×10^5)	11 060	> 100
+DEAE-dextran (Mol. Wt. 2×10^6)	2 250	20·9

* Average of five plates, 1/20 of the surface counted in each plate.

In contrast, incubation of the cells with the polycation DEAE-dextran under the same conditions decreases the plating efficiency by a factor of five (Table II).

Although more work is needed to establish this point, we have developed from these results the working hypothesis that it is the difference in charge between the negative sites of the cell surface and the charge of the network of polymers surrounding the cell which is relevant for the induction or repression of mitosis. In other words, one may say that sulphated polysaccharides act by contact inhibition. Another possibility is that they trap calcium ions bound to the negative sites of the cell surface, and hence modify the permeability to metabolites, the intracellular concentration of which might be critical for the induction of DNA replication.

(2) *Role of adenine nucleotides in cellular surface changes*

Cell surface changes are associated with transformation from stage *1* to stage *2*. They can be visualized with the ruthenium red technique first described by Martinez-Palomo and Brailovsky (1968). The surfaces of cells grown in monolayers and properly washed and fixed bind the dye ruthenium red. Transformed cells (at stage 2) bind more ruthenium red than cells at stage 1 or normal fibroblasts. Moreover, the thickness of the layer stained by the dye varies from place to place in the transformed cells, while only small variations could be seen in cells at stage 1. These differences remain after treatment with trypsin, although the stainable material becomes more compact (Torpier and Montagnier 1969).

The specificity of the fixation of ruthenium red is not known. It is generally assumed that it binds acidic groups of polysaccharides of the cell surface when they are in high concentration. An increased fixation would therefore mean either that the concentration of these components is increased at the surface of transformed cells, or alternatively that more sites are unmasked.

The important point is that the increase in binding only occurs if the cells have previously been grown in the presence of adenine nucleotides (Torpier and Montagnier 1970), the same requirement as for growth in agarose containing sulphated polysaccharides. In the absence of these nucleotides the growth of cells in monolayers is normal, although the cells are less refringent and more spread out on the supporting surface. The thickness of the layer associated with ruthenium red then becomes close to that of untransformed BHK cells, which remains the same whether nucleotides are present or not.

Another surface change depending on adenine derivatives has been detected in experiments using phospholipase C. We have shown (Montagnier 1970a) that the phospholipids of the plasma membrane are protected from the lytic action of this enzyme if the cells have been grown in the presence of BTP or adenine nucleotides for at least 24 hours before treatment. However, in this case, untransformed BHK behave similarly to their more transformed derivatives.

The combined results of these two investigations suggest that one or several components bearing acidic groups (glycoproteins, mucopolysaccharides) are assembled or synthesized in the presence of exogenous nucleotides in such a way that they protect the polar groups of phospholipids of the plasma membrane. The difference between cells at stages 1 and 2 may be that in the former, these components are always covered by more superficial components (proteins?) so that they are partially masked (Montagnier 1970a).

Since all the nucleosides and nucleotides of adenine are active, it is difficult to determine which of these components is actually involved. The fact that exogenous nucleotides are required suggests that the nucleotides of the internal pool do not freely enter the compartment where they act; perhaps this compartment is the cell surface itself. Since 5′- or 3′-AMP are as active as cyclic AMP, a specific role of the latter seems to be excluded.

Whatever the mechanism of these differences may be, it is of interest to note that by using the binding of plant agglutinins to glycoprotein sites of the cell surface, Burger (1969; see also this symposium, pp. 45–63) has come to a similar conclusion: namely that these sites are masked by proteins removable by trypsin in untransformed cells, and unmasked in virally transformed cells.

It seems therefore that a general feature of cells transformed by oncogenic viruses would be a deficiency in the synthesis or the attachment to the cell surface of superficial proteins. Two questions then have to be answered: (1) How are these changes related to the induction of DNA replication in abnormal conditions? And (2) what is the genetic determinism of these changes: are they directly or indirectly induced by the viral genome?

(1) On the first question, it seems likely that the nuclear membrane is a relay between signals coming from the cell surface and the triggering of DNA replication.

Each chromosome contains a certain number of replication units (Taylor 1968); if one extends the replicon theory to eukaryotic cells, one is led to assume that each of these replication units depends for initiation on sites bound to the nuclear membrane. The necessarily coordinated replication of these units would require that each of these sites has a common area responding to the same signals.

These signals may be micro- or macromolecular effectors coming from the external medium: in this case changes in the cell surface would result in an increased permeability to these signals. It is also possible that there is some kind of functional relationship between sites of the cell surface and sites of the nuclear membrane, so that a configurational change in the former could induce a similar change in the latter, and reciprocally.

(2) On the role of the viral genome, we have mentioned that the final stage of transformation (2a or 2b) depends on the transforming agent: cells at stage 1a become 2a after transformation by RSV (Schmidt-Ruppin or Bryan strain) or HSV (hamster sarcoma virus).

Transformation by a small plaque strain of polyoma virus produces a majority of clones at stage 2b. In contrast, most of the clones transformed by a large plaque strain of polyoma virus are only at 2a. This suggests that the genetic information of the virus controls the degree of transformation in some way (Montagnier 1970b).

However, the final stage of transformation induced by a given strain of virus also depends on the initial state of the cell: BHK cells at stage 1b always give cells at stage 2b, upon transformation by the large plaque strain as well as by the small plaque strain. Furthermore, cells at stage 2a can evolve spontaneously and progressively, following serial passages, to stage 2b.

These results are best interpreted by assuming that the viral genes controlling the properties described here act by triggering a stepwise derepression of some cellular genes, perhaps belonging to those controlling cell surface characteristics (Montagnier 1970c).

SUMMARY

The growth of BHK21 cells in an agarose gel depends on effectors, both promoters and inhibitors, present in the medium. Following spontaneous evolution or transformation by oncogenic viruses (RNA or DNA- containing), the cells have a modified capacity to respond to these effectors, while dependence on new effectors appears. From the behaviour of various clones towards these effectors, several degrees of transformation can be distinguished. The higher degrees of transformation are correlated with detectable alterations of the cell surface, also controlled by the new effectors.

REFERENCES

BURGER, M. M. (1969) Proc. natn. Acad. Sci. U.S.A. 62, 994–1001.
MACPHERSON, I. and STOKER, M. (1962) Virology 16, 147–151.
MARTINEZ-PALOMO, A. and BRAILOVSKY, C. (1968) Virology 34, 379–382.
MONTAGNIER, L. (1966) Path. Biol., Paris 14, 244–251.
MONTAGNIER, L. (1968) C. r. hebd. Séanc. Acad. Sci., Paris, Sér. D 267, 921–924.
MONTAGNIER, L. (1970a) In Défectivité, démasquage et stimulation des virus oncogènes, pp. 19–32. Paris: Centre National de la Recherche Scientifique.
MONTAGNIER, L. (1970b) Bull. Cancer 57, 13–22.
MONTAGNIER, L. (1970c) In IVth Symposium for Comparative Leukaemia Research, ed. Dutcher, R. M. Basle: Karger.
MONTAGNIER, L. and MACPHERSON, I. (1964) C. r. hebd. Séanc. Acad. Sci., Paris, Sér. D 258, 4171–4173.
MONTAGNIER, L., VIGIER, P., BOUÉ, J. and BOUÉ, A. (1967) C. r. hebd. Séanc. Acad. Sci., Paris, Sér. D 265, 2161–2164.
SANDERS, F. K. and SMITH, J. D. (1970) Nature, Lond. 227, 513–515.
STOKER, M., O'NEILL, C., BERRYMAN, S. and WAXMAN, V. (1968) Int. J. Cancer 3, 683–693.
TAYLOR, J. H. (1968) J. molec. Biol. 31, 579–594.
TORPIER, G. and MONTAGNIER, L. (1969) Ann. Inst. Pasteur Lille, 20, 203–210.
TORPIER, G. and MONTAGNIER, L. (1970) Int. J. Cancer 6, 529–535.

DISCUSSION

Pitts: Is the effect on BHK21 cells specific for adenine nucleotides?
Montagnier: No. We tried all four bases. We began by showing that

RNA itself was active, and then we found that ribonuclease-treated RNA was also active. All derivatives of adenine are active—adenosine, ADP, ATP. Some of the guanine derivatives are active, for example GMP, but none of the pyrimidine derivatives.

Dulbecco: Is anything known about adenyl cyclase in these cells?

Montagnier: Dr Bürk reported (1968) that adenyl cyclase activity was reduced in crude membrane preparations of polyoma-transformed BHK cells compared to untransformed BHK. Exogenous cyclic AMP has exactly the same effect as 3′- or 5′-AMP on the growth of BHK cells in the presence of sulphated polysaccharides and on the ruthenium red staining. Hypoxanthine is also active; theophylline is not.

Stoker: Did you say trypsin affects the ruthenium red staining of stage 1a?

Montagnier: After treatment with trypsin the ruthenium-stained layer looks more compact in all cases but is still much thinner in cells at stage 1a (untransformed) than in cells at stage 2a. However, after trypsinization, in some clones of transformed cells (particularly those transformed by hamster sarcoma virus) this layer is more labile, and can be detached, without altering the plasma membrane, just by incubating the cells in saline.

Dulbecco: Is insulin required for its hormonal activity? Would a hormonally inactive analogue work?

Montagnier: We used only crystalline insulin from two commercial sources.

Macpherson: The lowest effective concentration of crystalline insulin in our experiments was rather high (about 1 μg/ml), so the insulin may be acting in some way unrelated to its hormonal activity. We have not tested the effect of analogues.

Rubin: Herbert Morgan found that when he infected chick embryo cells with the non-transforming leucosis viruses the ruthenium red staining increased as it did when he infected them with transforming Rous sarcoma virus (Morgan 1968).

Montagnier: We have only looked at permanently transformed, non-virus-producing cells. Starting from the original BHK21 clones, cells transformed by Rous sarcoma virus did behave exactly like cells transformed by polyoma virus. As I remember Morgan's work, the increase of ruthenium red staining was rather small in cells infected with various types of Rous sarcoma virus or Rous avian virus, except for the Fujinami strain, which is known to induce a very large increase in the production of hyaluronic acid.

Rubin: The greatest increase was certainly found with the Fujinami strain of RSV, but an increase was found even with the non-transforming viruses. The main point is that there was no correlation with the loss of contact inhibition.

Montagnier: As I said, we don't know how specific the ruthenium red staining is. In Morgan's cells, it seemed to stain the acid polysaccharides of the hyaluronic acid type. It is not as simple in our system.

Warren: Morgan (1968) found that treating the cells with hyaluronidase removed much of the material that picks up ruthenium red stain. Have you tried this and other degradative enzymes such as neuraminidase, trypsin or phospholipases?

Montagnier: In our cells, hyaluronidase was not effective, nor was neuraminidase, at low doses; at high doses it lysed the cells. Phospholipase C seems to have some effect on the lability of the cell layer but this effect was difficult to reproduce.

Warren: Could it be that the adenine bases you are adding simply have to do with the condensation of material on the surface of the cells, almost a chemical effect rather than an effect on the metabolism of the cell?

Montagnier: No, because some metabolism is required for the effect to be observed; the cells have to be growing. If you just incubate the cells with adenine nucleotides for one hour, this is not sufficient to observe a detectable increase of the layer or to change the resistance to phospholipase C.

Warren: I'm not saying that the material isn't biosynthesized; only that the effect might be due to precipitation of adenine or any of the other bases that you add.

Montagnier: One cannot rule this out, although it seems unlikely to me because one needs at least 24 hours—that is to say, more than a full cell cycle—to get the changes.

Bergel: I wonder if you would observe differences in the electrophoretic behaviour, and thus changes in the charge, of your cells after treatment with trypsin or phospholipase? This, so one is told, is due to the charge effect of exposed neuraminic or sialic acid.

Montagnier: The situation about sialic acid residues in transformed cells and their role in changes of surface charge is very confused; some authors have found an increase, others a decrease in transformed cells.

Macpherson: Polyoma-transformed BHK21 clones have a wider scatter of electrophoretic mobilities (EPM) but the same mean as their BHK21 precursors. Treatment with neuraminidase reduced the mean EPM of both normal and transformed cells by about the same amount and also narrowed the scatter of EPM in transformed cells so that they now resembled the normal cells (Forrester, Ambrose and Macpherson 1962).

Clarke: Is there a quantitative relationship between the dextran sulphate content of the medium and the adenine requirement which distinguishes between groups *2a* and *2b*?

Montagnier: If one increases the doses of dextran sulphate (to more than 15 μg/ml), it becomes inhibitory in any conditions. At these doses, the shape of the dose–response curve of colony formation showed some variations from one clone to another, which I could not relate to my classification into *a* and *b*. Under these conditions, the adenine requirement is no longer a limiting factor.

Pitts: You say a complete cell cycle is necessary before adenine shows its effect. Have you tried using analogues of adenine which would prevent the cell cycle?

Montagnier: I have tried several groups of analogues; there are no definite results yet.

Rubin: Our chicken cells also require adenine for growth, and we find immediately on adding adenine that thymidine uptake into the soluble pool increases considerably.

Cunningham: We have not examined the effect of adenine or adenosine on the transport of other substrates by 3T3 cells. However, our studies have shown that transport of adenosine, unlike the other nucleosides, does not increase during the 30-minute interval after adding fresh serum to confluent 3T3 cells.

Dulbecco: With regard to requirements for serine and adenine, can you exclude interference from contamination with mycoplasma?

Montagnier: One can never exclude a latent contamination of this type. However, (1) all our culture media contained aureomycin; (2) weekly controls by electron microscopy never showed mycoplasma in any clone; (3) the requirement for serine appears immediately after transformation; and (4) the viral preparations were irradiated before use with 20 000 R (γ rays) in order to inactivate any bacterial contaminant.

Dulbecco: It is essential in many cases to show the absence of mycoplasma; a good way to detect it is to give a pulse of thymidine and look for grains outside the nucleus in radioautographs. I have detected abundant mycoplasma by this means where it was not detected by culturing in experienced laboratories.

REFERENCES

BÜRK, R. R. (1968) *Nature, Lond.* **219,** 1272–1276.
FORRESTER, J. A., AMBROSE, E. J. and MACPHERSON, I. A. (1962) *Nature, Lond.* **196,** 1068.
MORGAN, H. R. (1968) *J. Virol.* **2,** 1133–1146.

THE SIGNIFICANCE OF SURFACE STRUCTURE CHANGES FOR GROWTH CONTROL UNDER CROWDED CONDITIONS

MAX M. BURGER

Department of Biochemical Sciences, Princeton University, Princeton, New Jersey

NORMAL fibroblasts in tissue culture stop growing at the point where they reach the cell density of a confluent monolayer. These cells are therefore sensitive to some sort of growth control, the mechanism of which is so far unknown. Under identical nutritional conditions, virally as well as chemically transformed cells grow past the confluent monolayer stage and by piling up eventually reach 5–10-fold higher saturation densities.

Even though this phenomenon, called contact inhibition of growth or density dependent inhibition of growth (Stoker and Rubin 1967), is dependent on factors in the growth medium, normal cells clearly display a different sensitivity to these growth factors from tumour cells.

Since tumour cells show growth control of this type which is apparently dependent on the close proximity of cells, and since other properties of tumour cells, like metastasizing and adhesion (Coman 1944; Abercrombie and Ambrose 1962), may also be related to the cell surface, it had been suggested some time ago (Pardee 1964) that tumour cells may have modified surface membranes which are responsible for their altered growth behaviour. In the following, such a change in transformed cells is briefly described (A) and correlations between the change and the growth behaviour of various cell lines are reported (B). Furthermore, experiments will be discussed which indicate that infection by oncogenic viruses gives rise to the same change (C) and that during a certain stage in the cell cycle (mitosis) all cells undergo briefly the same surface change, after which the normal surface appearance is reinstalled again for the rest of the cycle (D). Finally, it will be shown that normal cells can escape density dependent inhibition of growth if their cell surface is artificially modified to the type seen permanently in transformed cells (E).

A. AN ARCHITECTURAL CHANGE IN TRANSFORMED SURFACE MEMBRANES

Cells transformed by polyoma, SV40 and adenovirus have been found to display surface receptor sites for an agglutinin isolated and purified from wheat germ lipase (Burger and Goldberg 1967; Burger 1970). The normal parent cell lines did not agglutinate under identical conditions even though 5–15-fold higher concentrations could eventually bring about agglutination as well. The surface determinant group contains the carbohydrate N-acetylglucosamine, presumably the disaccharide. Isolation of the receptor site and partial purification indicates that it is a surface glycoprotein, although some glycolipids may also have receptor function for the agglutinin (Hakomori and Murakami 1968). Several aspects of this work have been confirmed by various laboratories (Uhlenbruck, Pardoe and Bird 1968; Biddle, Cronin and Sanders 1970).

Normal cells seem to contain a similar receptor, from isolation (Burger 1968) and immunological quantification studies. A detailed chemical analysis of the receptors from the two cell types is still required however and until such a study is made, differences can by no means be clearly excluded.

These results led us to investigate how the receptor sites apparently also present in normal cells could be exposed and how their existence could unambiguously be demonstrated by agglutination. When normal tissue culture cells were exposed briefly to various degradative enzymes, only proteases were found to expose the same agglutination site as found on untreated transformed cells (Burger 1969). Lipases, glycosidases or mucopolysaccharidases had no effect. Since we have recently found that under appropriate conditions, less than 0·002 per cent pronase could bring about this exposure in 3T3 mouse fibroblasts in 30 seconds, we hardly expected the release of a large proteinaceous cover layer, and in fact were unable to isolate a surface protein specific for normal cells that would have been cleaved off. We are continuing to search for such a cover protein, however, using more refined procedures. For the present we have to leave the question open whether the proteolytic conditions that bring about an exposure of the same agglutinin receptors permanently exposed in transformed cells do so by releasing a cover protein or just by a rearrangement in the surface membrane caused by the cleavage of some surface peptide bonds (Burger 1970).

Recent studies from Sachs' laboratory (Inbar and Sachs 1969a, b) have shown the exposure of receptor sites for another agglutinin (concanavalin A) and have in general confirmed the studies described for wheat germ agglutinin. In the meantime, we were able to add a few more agglutinins to this list. During such screening work with new agglutinins, preliminary results began to indicate that while the majority of agglutinin receptor sites seem to become exposed in the course of transformation as well as after proteolytic treatment, some agglutinins and some specific cell lines do not behave in this way. In certain cases (Sela *et al.* 1970) transformed cells agglutinated less, and in certain cases protease-treated cells agglutinated less than untreated normal cells (Burger, unpublished observation). A possible interpretation of such findings is given in Fig. 1, and suggests a "swivel" mechanism.

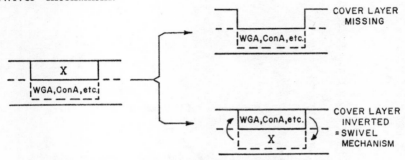

FIG. 1. Two possible mechanisms simultaneously explaining the appearance of new sites by exposure (wheat germ agglutinin site, WGA, and concanavalin A site, Con A) and the disappearance of sites previously present in normal cells (site X stands for new sites, now under investigation).

For a survey of all theoretically possible mechanisms, see Burger (1970). The disappearance of antigenic sites as well as the appearance of new antigenic sites have both been reported in the literature on tumours for over 30 years but they have rarely been conceived as a manifestation of the same surface process. However, we do not consider the two possible mechanisms illustrated here to be unique and an explanation of all antigenic changes in tumour cells.

The disappearance of a given site (in this case site X) after treatment with proteolytic enzymes could be due to either of the two mechanisms, and to decide which is operating in a particular case the hidden presence of the antigen after proteolytic treatment has to be shown by isolating the site and estimating it quantitatively (Jansons, Sakamoto and Burger 1970).

B. CORRELATION BETWEEN THE DEGREE OF SITE EXPOSURE AND OF GROWTH AFTER CONFLUENCY

A comparison of cell lines that display various degrees of density dependent inhibition of growth (DDI) has shown that the degree to which the wheat germ agglutinin receptor site occurs in an exposed form is correlated with the degree to which DDI is lost (Burger 1970). At the present time we do not think that the wheat germ agglutinin site *per se* has to be involved in growth regulatory phenomena but that the general architectural membrane change which we have found, as manifested by the wheat germ agglutination reaction, may give rise to the growth control changes. The question of whether it is the absence of the receptor cover or the exposure of the receptor layer which is the cause of the change in growth control is still an open one.

More convincing support for a relation between DDI of growth and site exposure comes from a comparison, done in collaboration with Dr R. E. Pollack, between various cell lines derived from a single stem line

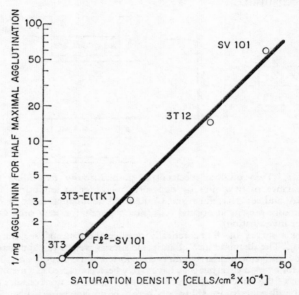

FIG. 2. Correlation between loss of density dependent inhibition of growth and agglutinability. These results were given earlier in non-graphic form by Pollack and Burger (1969).

(Pollack and Burger 1969). Here various virally (SV 101) as well as spontaneously (3T12) transformed lines with extensive and intermediate losses of DDI of growth turned out to have the agglutinin receptor exposed to a corresponding degree (Fig. 2). The best support comes from a cell line which Pollack, Green and Todaro (1968) selected from a virally transformed line for recovery of DDI of growth. This variant (Fl^2-SV101), sometimes erroneously called a revertant, has regained almost complete DDI of growth and has covered up the agglutinin receptor again (Pollack and Burger 1969). The fact that this same cell line still contains the tumour virus genome from which transcription of its own RNA still proceeds and from which specific protein (T antigen) is translated, supports very strongly the argument that at least covering, if not also exposure of the agglutinating receptor, is a host function which is tightly coupled with the expression of malignancy in the tumour cell. This result also implies that the presence of a tumour virus genome in animal cells that can be transformed is not a sufficient condition, although probably a necessary one, for the exposure of the agglutinin site or for transformation. Inbar, Rabinowitz and Sachs (1969) have shown that another variant line which has also regained its DDI of growth, although it was isolated by an entirely different procedure, had lost its agglutinability for yet another agglutinin.

C. EXPOSURE OF SITES DURING INFECTION BY ONCOGENIC VIRUSES

All oncogenic viruses giving rise to lytic infection that we have tested so far have exposed the wheat germ agglutinin receptor site, usually as a late function (Sheppard, Levine and Burger 1971). In those cases where transformation occurred eventually in only a small fraction of cells but where all cells had been infected initially it turned out that at the beginning almost all cells exposed this site transiently (Sheppard, Levine and Burger 1971; Ben-Bassat, Inbar and Sachs 1970), an observation Stoker had already made earlier (personal communication 1969) and which possibly belongs to the phenomena designated "abortive infection".

If the virus were simply acting as a co-repressor in the transformation process and were thereby suppressing the formation of the trypsin-sensitive cover layer, a simple experiment should be able to support such a mechanism of action. Uninfected normal cells that are briefly treated with protease and consequently become agglutinable lose their agglutinability in about 6 hours, since the cover components are repaired again, a process which incidentally can be inhibited with cycloheximide. When normal cells were then infected and treated with protease at various times after infection, repair could be demonstrated to take place right up to the time when exposure of the site occurs normally in the course of infection (Sheppard, Levine and Burger, unpublished observation). Lack of regeneration of the cover site after infection would have been quite a strong indication of a co-repressor role of the transforming virus. Our results make such a role unlikely, and leave several other possibilities open. The most obvious one is the actual destruction of the cover site by degradative enzymes acting as a consequence of viral transformation. Such a hypothesis seems to us compatible with the fact that proteolytic enzymes added from the outside achieved the same site exposure as did viral transformation.

Studies on two different types of polyoma virus mutants indicate that a specific gene(s) seems to be required for exposure of the agglutination receptors as well as for bringing about the transformed state.

A polyoma virus mutant has been isolated by Benjamin (1970) which is only able to infect lytically mouse fibroblasts that have already been transformed, but not untransformed cells. This virus mutant apparently lacks one genetic function that is provided by the transformed cell for successfully infecting mouse fibroblasts. Table I shows that the mutant

TABLE I

ABSENCE OF THE SURFACE CHANGE IN CELLS INFECTED WITH MUTANT VIRUS

Cells	Agglutinin necessary for half-maximal agglutination	
	Wheat germ agglutinin $\mu g/ml$	Concanavalin A $\mu g/ml$
3T3*	300–500	600
3T3, polyoma virus-transformed	40	120
3T3 infected by polyoma virus mutant NG-18	200–300	600
NG-23	200–300	
NG-59	200–300	
HA-33	200–300	
3T3 infected by wild-type polyoma virus	20–40	150

*Well contact-inhibited 3T3 cells sometimes require up to 800 μg/ml.

Half-maximal agglutination concentrations for the two agglutinins were described in Benjamin and Burger (1970), from where the contents of this table were taken.

virus has also lost the genetic function of exposing the agglutinin receptor layer even though other functions were still retained (Benjamin and Burger 1970).

Studies on temperature-sensitive polyoma virus mutants (Eckhart, Dulbecco and Burger 1971) indicate that the expression of the surface change as measured by exposure of the agglutination sites is affected at non-permissive temperatures. Figure 3 shows the appearance of the agglutination site under permissive conditions for the virus (32° C) and its disappearance under non-permissive conditions (39° C). As will be discussed by Dr Dulbecco, a "shift-up" to non-permissive temperatures of an already transformed cell covered up the receptor site, while a "shift-down" exposed it again (Eckhart, Dulbecco and Burger 1971). It can be seen in Fig. 3 that disappearance of the site occurred earlier than exposure. This is presumably due to regeneration of the surface membrane which in normal cells after trypsin treatment takes just about the same time, approximately 6 hours, as already mentioned. Exposure, however, seems to take about the time of a full cell cycle, an observation which might be explained by the results discussed under D. Other explanations cannot yet be excluded at this point.

Evidence begins to accumulate that inhibition of synthesis of host DNA immediately after infection of a tissue culture with an oncogenic virus prevents the exposure of agglutinin receptor sites. In three different efforts, this could be shown with 5-fluorodeoxyuridine (FUdR) (Benjamin and Burger 1970) and hydroxyurea (Eckhart, Dulbecco and Burger 1971) on polyoma virus-infected 3T3 mouse fibroblasts and with FUdR on SV40 and adenovirus-infected CV-1 monkey cells (Sheppard, Levine and Burger 1971). With additional evidence published earlier (Benjamin and Burger 1970; Eckhart, Dulbecco and Burger 1971; Sheppard, Levine and Burger 1971), the preliminary conclusion could be drawn from this work that it is

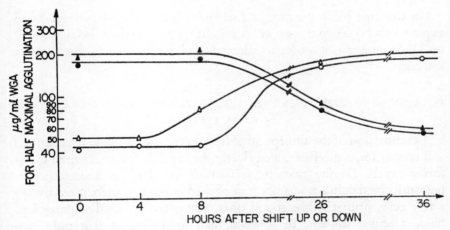

FIG. 3. Appearance and disappearance of surface alteration after shift in temperature of ts-3-transformed BHK cells.

▲—▲, clone 7-C ⎫
●—●, clone 1 ⎬ shifted from 30° to 32°C.

△—△, clone 1 ⎫
○—○, clone 7-C ⎬ shifted from 32° to 39°C.

host and not viral DNA synthesis which is required and that this synthesis of host DNA is presumably a necessary but not a sufficient requirement for site exposure (Benjamin and Burger 1970). Ben–Bassat, Inbar and Sachs (1970) have suggested that host cells have to go through one round of cell division in order to express the property of site exposure. In the light of our finding that all normal cells expose agglutination sites briefly during mitosis (see next section), we should like to offer the following two hypotheses for all these findings:

(1) Transformation and abortive transformation require host DNA synthesis in order to establish the surface membrane changes detected by the exposure of agglutinin receptor sites.

(2) Surface changes observed in transformation and abortive transformation occur earliest at the first mitosis after infection and may be due to fixation of that normal mitotic surface pattern in all subsequent interphases. In those cases of lytic infection where the cells do not go through a round of cell division between infection and virus formation but where the agglutinin receptors are nevertheless uncovered, one has to assume that only those host genes are expressed which are necessary for exposure, while all the others which are also involved in the mitotic process would not be expressed. This type of block, or some block at the translational level, may lead to an abortive mitosis concerning only the cell surface.

For the time being we cannot distinguish between explanation (1) and explanation (2), since the second possibility is presumably also dependent on DNA synthesis in order to allow the cell to proceed to mitosis after infection.

D. EXPOSURE OF SITES DURING THE MITOTIC PROCESS IN THE CELL CYCLE OF NORMAL CELLS

A comparison of the adsorption of fluorescent agglutinin to transformed and untransformed cells showed clearly the superior fluorescence of transformed cells. During that study it struck me that the fluorescence in untransformed cultures was not homogeneous and that there were always a few cells, primarily the round ones, that were brilliantly fluorescing. Since a senior student, R. Johnson, had found earlier that polyoma-transformed 3T3 cells adsorbed just about 20 times more ^{3}H-labelled wheat germ agglutinin per unit surface area than normal 3T3 cells (Johnson 1968) and since mitosis took about 1 hour in a total cell cycle of 20 hours, Mr Fox, Dr Sheppard and I (1971) made a careful analysis of the cell cycle.

If permeability barriers are not damaged by fixation, those cells in mitosis will adsorb fluorescein-labelled agglutinin specifically. The same hapten inhibitors prevented adsorption as were earlier found to inhibit agglutination (N-acetylglucosamine and di-N-acetyl chitobiose), while control sugars like glucose had no effect. Fluorescein-labelled bovine serum albumin adsorbed poorly and certainly not preferentially to mitotic cells.

It seemed that primarily metaphase, anaphase and telophase cells fluoresced (Fig. 4) but hardly any cells in prophase. On the other hand, exposure of the agglutinin receptor sites and adsorption of the agglutinin continued to persist to some degree into early G1.

Mitotic and fluorescence indices corresponded quite well in synchronized cultures (see Fig. 5), regardless of whether synchrony was achieved by

Fig. 4. 3T3 cells after adsorption of fluorescein isothiocyanate-labelled agglutinin.

    (*a*) With UV illumination (metaphase).
    (*b*) The same in dark field (metaphase).
    (*c*) With UV illumination (anaphase).
    (*d*) The same in dark field (anaphase).
    (*e*) With UV illumination (late telophase).
    (*f*) The same in dark field (late telophase).

In order to demonstrate that these cells are mitotic cells on a photograph, the cells were first fixed under controlled conditions in ethanol, then exposed to agglutinin labelled with fluorescein isothiocyanate, counterstained with Evan's blue and mounted in Elvanol. Only with this procedure, devised to allow some unspecific uptake of the fluorescent stain into the cytoplasm, could the nuclei be visualized as dark zones which do not contain fluorescent material. That most of the fluorescence was still due to specifically and presumably surface-adsorbed agglutinin could be shown by a decrease of 70 per cent in the fluorescence after addition of the hapten inhibitor N-acetylglucosamine. For regular observations, the cells were first exposed to fluorescein isothiocyanate-labelled agglutinin and then fixed or in other cases not fixed at all.

[*To face page 52*

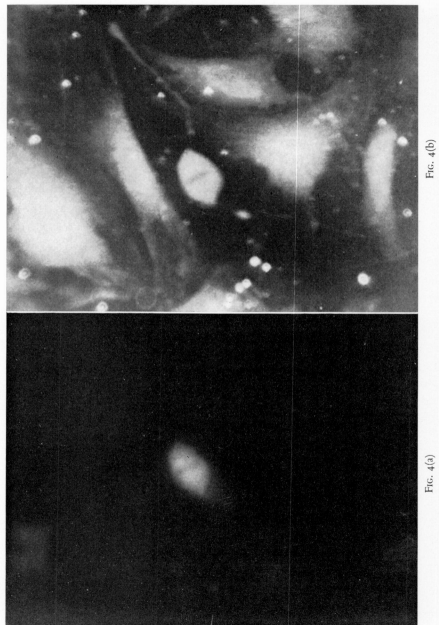

FIG. 4(a)

FIG. 4(b)

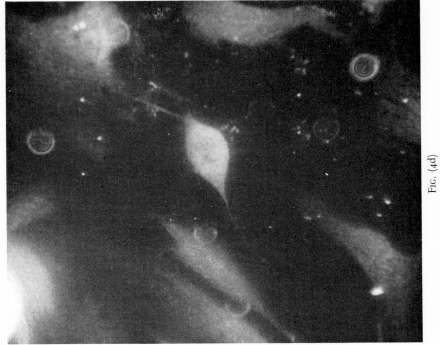

Fig. (4d)

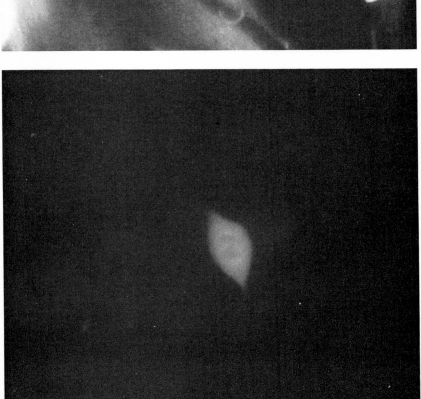

Fig. 4(c)

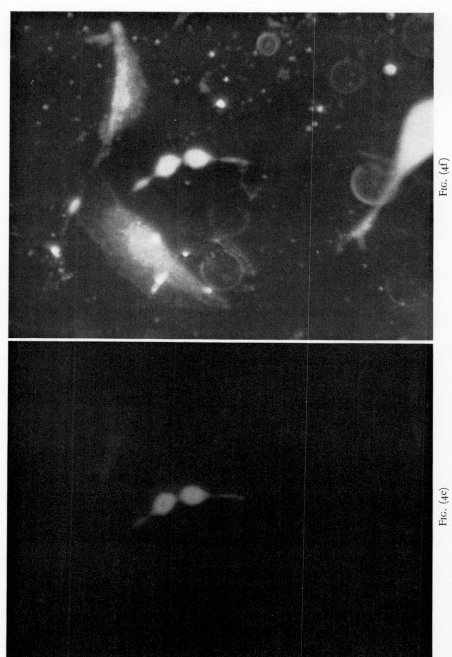

FIG. (4f)

FIG. (4e)

thymidine block, by 50 per cent serum added to a confluent culture or by subculturing with trypsin.

DNA synthesis seems to be required after viral infection to keep this same surface site exposed, as discussed in the previous section. DNA synthesis is generally assumed, however, also to be required in the normal cell cycle for M phase to occur.

From this, together with our data, we should like to suggest two testable hypotheses:

(1) Lytic infection, abortive transformation and transformation differ in their regulation from the normal cell cycle in that the cell is no longer able to cover the sites exposed during mitosis, and the suggestion is that it essentially "freezes" in the mitotic conformation.

(2) For the normal cell cycle we propose the very simplistic but useful model shown in Fig. 6. Nuclear events, such as DNA synthesis, would lead to a mitotic cell surface configuration, among other events typical of mitosis. The mitotic cell membrane, or processes involved in making it, on the other hand, would become a required signal for phases later on in the cell cycle, like S or G1 and G2.

E. PROTEOLYTIC ENZYMES INITIATE CELL DIVISION AND ESCAPE FROM DENSITY DEPENDENT INHIBITION OF GROWTH

Once it was shown that small amounts of proteolytic enzymes could bring about a change in the surface structure that was similar to that brought about by transformation, the obvious question arose whether this surface change could be of significance to the growth behaviour of the cell. In other words, a brief proteolytic treatment of the surface of a normal cell may not only convert the surface structure to the type seen permanently in transformed cells, but may bring about a transient escape from density dependent inhibition of growth. In view of the phenotypic nature of this change, the cell should repair its surface again after the removal of the proteolytic enzyme and eventually should return to its previous state in which it is susceptible to DDI of growth.

Preliminary results from experiments done three years ago convinced us further of the validity of this working hypothesis. During a study on interactions between transformed and normal cells in tissue culture, stimulating effects of transformed cells upon confluent 3T3 mouse fibroblasts were noticed. The best results were achieved when L1210-leukaemia cells were used as the stimulating cells, since they could be layered over the 3T3 cells and removed fairly easily after a given time. They were also easily

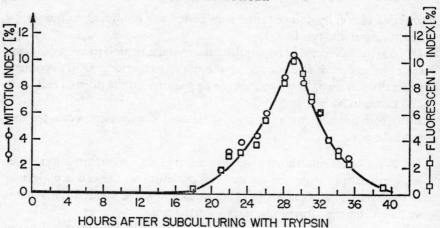

FIG. 5. Mitotic and fluorescence indices of synchronized 3T3 cells. Cells were synchronized by trypsinizing a confluent 3-day-old mono-layer of 3T3 cells and replating at lower densities on coverslips. The maximum percentage of synchrony obtained (12–15 per cent) is approximately that expected from other techniques. Coverslips were exposed to fluorescein isothiocyanate-labelled agglutinin, fixed with ethanol, stained with Evan's blue and mounted in Elvanol. In control experiments, cells were exposed to fluorescein-labelled agglutinin and counted without fixing and staining, and fluorescence indices were identical to those reported in the figure. Blind counts of several hundred cells were made by two investigators and were in good agreement.

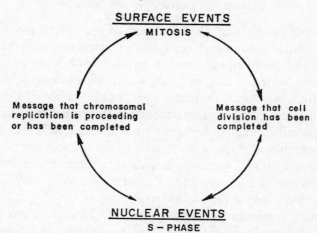

FIG. 6. Hypothetical model for a function of the mitotic cell surface change in the cell cycle.

distinguished on morphological grounds (primarily size) from 3T3 fibro-blasts. Figure 7 shows that the growth response was dependent on the number of leukaemia cells added and Fig. 8 that an exposure interval of

3 hours was sufficient to bring about the escape from density dependent inhibition of growth. Culture medium from L1210 cells grown in tissue culture gave only a very poor response (10–20 per cent higher saturation density), while ascitic fluid from mice inoculated intraperitoneally with L1210 cells resulted in about a doubling of the saturation density. Since tissue culture medium from L1210 cells contained only very little proteolytic activity while ascites fluid had 5–10-fold more, we thought that the

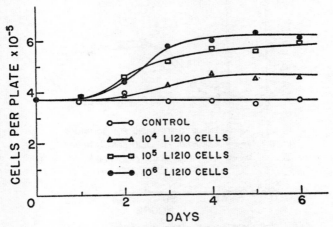

FIG. 7. Proportional growth stimulation of 3T3 cells by exposure to increasing numbers of L1210 leukaemia cells. Various numbers of L1210 leukaemia cells were added to 3T3 mouse fibroblasts for 13 hours in conditioned medium. L1210 cells were removed by pouring them off and rinsing with dilute EDTA solutions in Ca^{++} Mg^{++}-free phosphate buffered saline. Controls were treated identically. The controls received the same conditioned medium, which contained 3 per cent calf serum and had been depleted for 3 days on a 3T3 culture. Saturation densities were determined by counting cells released with 0·01 per cent trypsin in a Levy-Hausser chamber. At all three numbers of leukaemia cells added, less than one cell in a hundred 3T3 cells was seen in the counting chamber which had remained on the plates after rinsing.

growth stimulatory effect could be due to proteolytic enzymes. A recent more detailed study by Mr Robert Remo indicates that the growth stimulatory effect is, however, not due to proteolytic enzymes secreted into the medium by the transformed cell but rather by proteolytic enzymes located in the surface membrane and acting when the leukaemic cell makes contact with the untransformed 3T3 fibroblast (see legend to Fig. 8).

These results, together with the structural changes in surface membranes after brief proteolytic treatment, prompted us to test the working hypothesis that treatment with a low concentration of protease would stimulate the growth of cells in a density dependent state of growth inhibition.

The addition of large concentrations of trypsin was known to disperse the cells and subsequent growth in new medium is of course the expected course of events. When we harvested cells with EDTA instead of trypsin and then incubated them in conditioned medium containing a low concentration of serum at densities higher than confluency the cells removed with trypsin turned out to have reached a saturation density two to three

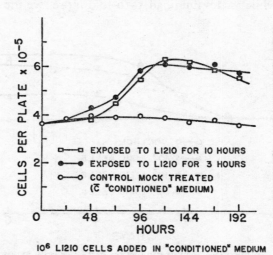

106 L1210 CELLS ADDED IN "CONDITIONED" MEDIUM

FIG. 8. Dependence of growth stimulation by leukaemic cells on time of exposure. The same conditions were used as described in Fig. 7. Incubations shorter than three hours have not yet been done. No correlation between the amount of proteolytic enzyme secreted by L1210 cells and the growth response to L1210 cell exposure has been found so far. Furthermore, the amount of proteolytic activity secreted during 3 hours is below the minimal concentration required for growth stimulation, as determined with the most active commercially available protease. Since we found that proteolytic inhibitors (TAME, TPCK) prevented the stimulatory effect of L1210 and since L1210 membrane preparations contain proteolytic activity and can stimulate growth of 3T3 cells, a possible interpretation of these results may be that L1210 cells stimulate growth of 3T3 fibroblasts by means of surface-located proteases when the cells make contact.

times higher than the cells removed with EDTA. We would like to interpret this result as another manifestation of the growth-stimulatory effect of trypsin, described below. On the other hand, we cannot exclude yet that during its removal EDTA has damaged cells more than trypsin did.

Initial experiments where low concentrations of trypsin were added to confluent 3T3 fibroblast cultures were not successful since the concentrations of serum necessary for obtaining growth were high enough to contain sufficient trypsin inhibitor to inactivate all the trypsin that was required to bring about the change in the surface architecture seen in transformed cells. Pronase and ficin, two proteases not from animal sources, were scarcely

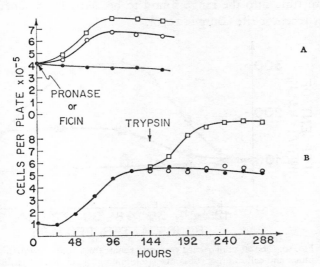

FIG. 9. Growth stimulation by proteolytic enzymes. A (*Upper graph*). 3T3 mouse fibroblasts were grown on 3·5 cm Falcon dishes in Dulbecco-Vogt medium with 3 per cent serum to saturation density and then pronase (Calbiochem) was added to the same medium to a final concentration of 5µg/ml for 45 minutes. Incubation medium was removed from both experimental and control dishes and the cultures were rinsed three times with 2 ml of medium drawn off from 2-day-old cultures (conditioned medium) to prevent a stimulatory effect of fresh serum. The same conditioned medium was used for continuous incubation of the treated and the control cultures. Daily thereafter the cell content per plate was counted after release with EDTA or 0·01 per cent trypsin, using a Levy-Hausser chamber since the Coulter counter did not give as reliable results. One point corresponds to 6 different experiments. o—o, pronase. ●—●, control. □—□, 0·0005 per cent ficin.

B (*Lower graph*). 3T3 cells were grown as in A except that the plates were inoculated heavily at day 0 rather than three days previously. At the arrow, a solution of sterile trypsin in phosphate buffered saline was added to bring the trypsin concentration to 0·007 per cent and phosphate buffered saline was added to the control. In general, Calbiochem trypsin (chymotrypsin-free) was used, occasionally also trypsin from Sigma (2 × crystallized) and from Grand Island Biological. Sometimes the effect of trypsin varied according to the source of the trypsin as well as the source and the condition of the serum used in the medium (serum trypsin inhibitor!). No such variation was found after incubations with pronase or ficin. □—□, trypsin. o—o, control with 0·1 per cent diisopropylfluorophosphate-inactivated trypsin (Worthington). ●—●, control without protease.

inhibited by serum and were therefore used first. The upper graph in Fig. 9 demonstrates that after a 45-minute exposure to these two enzymes, 3T3 cells go through just one round of cell division and level off 3–4 days later.

Mr K. Noonan has recently shown that the stimulating effect of pronase is exactly as pronounced after a five-minute exposure, which brings the exposure time into the range found to be necessary to uncover the agglutinin receptor site (Burger 1970).

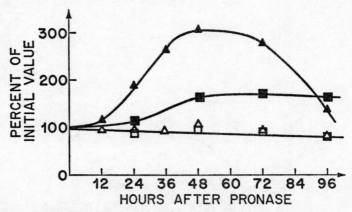

FIG. 10. The effect of 0·0001 per cent pronase for 2 hours on the mitotic index and cell numbers of contact-inhibited 3T3 cultures. Conditions were as in Fig. 7 except that pronase was present for 2 hours. Stimulation of growth (■—■) is given as percentage above control (□—□); stimulation of mitotic index (▲—▲) is also given as percentage above control (△—△).

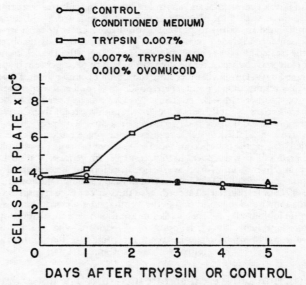

FIG. 11. Inhibition of the growth-stimulating effect of trypsin by ovomucoid. Conditions as in Fig. 7 except for ovomucoid additions as indicated.

The same stimulation of growth could be obtained with trypsin if the cells were kept in 4·8 per cent chicken and 0·2 per cent calf serum. The results were not, however, as reliable as when the cells were exposed to trypsin in phosphate-buffered saline or Dulbecco–Vogt medium and thereafter grown in medium containing serum (see Fig. 9, lower graph).

In a growth-stimulatory effect, one would expect both cell numbers and the mitotic index to increase, reflecting the increased amount of cell division. This is shown in Fig. 10. There seems to be nothing unique about the cells

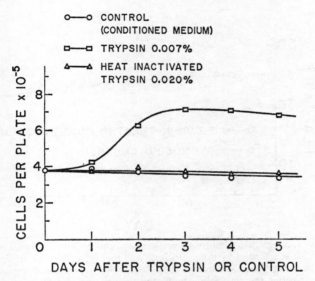

FIG. 12. Loss of growth-stimulatory effect after heat inactivation of trypsin. Conditions as in Fig. 7. Loss of proteolytic activity after heat treatment was measured, and amounted to over 99·8 per cent.

responding to trypsin as a growth stimulant since Rubin observed the same effect on chick embryo cells (Rubin 1970).

The effect of trypsin is due to its proteolytic activity, and not just to adsorption of the enzyme to the cell surface, as can be seen from the following experiments:

(1) Trypsin inactivated by diisopropylfluorophosphate is inactive (Fig. 9).

(2) The trypsin inhibitor ovomucoid prevents the growth-stimulatory effect while ovomucoid alone in a control experiment has no effect (Fig. 11).

(3) Heat-inactivated trypsin does not have the growth-stimulatory effect (Fig. 12).

(4) The growth-stimulatory effect of trypsin could be overcome with massive doses of serum from conditioned medium containing trypsin inhibitory activity.

With the evidence given so far, proteolytic action on the surface membrane is not the only possibility left. The proteases could have exerted their effect on a component in the medium or, in the experiment of Fig. 9 (lower), on a component of the medium that had previously attached to the surface. This criticism can be excluded by the experiment described in

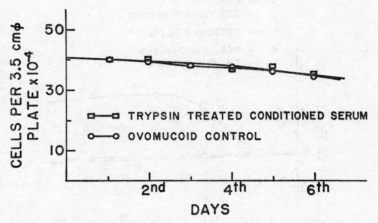

FIG. 13. Is growth stimulation mediated by a component in the medium which is activated by the proteolytic enzyme? Conditioned medium containing serum was treated for 2 hours with 0·01 per cent chymotrypsin-free trypsin (Calbiochem) and then the reaction was stopped with a two-fold excess of ovomucoid. Controls without trypsin but with ovomucoid and controls without trypsin or ovomucoid were run. Another control, where medium incubated alone for 45 minutes was added together with fresh trypsin to a confluent culture, did not show any impairment of the growth stimulatory effect.

Fig. 13, where no growth-stimulatory activity at all could be detected using complete medium treated with trypsin.

The criticism may be made that stimulation of growth could have been triggered within the cell after uptake of the protease. Weissmann and co-workers (1968) have indeed suggested that proteases may get into the nucleus and release repressors, in this case presumably those involved in growth control. Our findings point rather to a proteolytic effect upon membranes and we have formulated such a working hypothesis (Burger 1969, 1970). The growth-stimulatory effect of proteases described has supported this notion further, since trypsin attached to beads was able to bring about escape from the density dependent inhibition of growth as effectively as soluble trypsin. This enzyme in bead form could be washed away from the cell culture without leaving even traces, as could be confirmed with ^{3}H-labelled trypsin beads. Phagocytosis of the beads can thus be ruled out. Furthermore, incubation of the beads with serum did not release any trypsin previously attached to the beads. It can therefore be concluded that a brief treatment of the surface membrane is able to send cells that are contact inhibited and resting into a round of cell division.

The detailed molecular mechanisms of this effect are at present mostly speculative. In the following, a few of the possibilities are outlined:

(1) The surface changes caused by a protease may prevent the exchange of the natural message for contact inhibition.

(2) Both in transformation as well as after treatment with protease, rearrangement of the cell surface may cause an increased uptake of general nutrients (Pardee 1964; Cunningham and Pardee 1969) or specific growth factors from serum (Holley and Kiernan 1968).

(3) Even though, under the conditions used, no visible rounding up or retraction of pseudopods could be observed, cells may become less firmly attached to their substratum and particularly to neighbouring cells. This may provide them with the necessary mobility for cytokinesis, leading to division.

(4) Each cell may produce proteases which, if it is in contact with neighbouring cells, will trigger changes that will forward the message for cell division to the inside of those neighbouring cells. Information carriers might be of the type seen in the 3', 5'-adenosine monophosphate (cyclic AMP) system or cyclic AMP itself, or they might consist of an intricate series of structural and enzymic reactions, as found in complement fixation and lysis, where a cascade of proteolytic processes is touched off, starting on the surface membrane.

CONCLUSIONS AND SUMMARY

(1) A difference in the molecular surface architecture of transformed and untransformed plasma membranes has been described. It consists in the exposure of a series of agglutinin receptor sites in transformed cells. A similar change can be brought about transiently by briefly exposing normal cells to low concentrations of proteolytic enzymes. Together with these agglutinin receptors, surface antigens which were previously believed to be virus-specific and virus-coded can also be exposed in untransformed cells after protease treatment (Häyry and Defendi 1970), indicating that a whole layer of macromolecules must be exposed, some of them even specifically by specific viruses.

3*

(2) When transformed cell lines which have lost a certain growth control mechanism (density dependent inhibition of growth) to different degrees were compared, a correlation could be found with the degree of exposure of these agglutinin receptor sites.

(3) Brief treatment with low concentrations of proteolytic enzymes can bring about not only the same change in normal cells as is observed in transformed cells but also the same loss of density dependent growth control as is observed in transformed cells.

(4) The sites that are permanently exposed on the surfaces of transformed cells seem to be exposed in normal cells also, but only during a short interval of the cell cycle, namely mitosis. After mitosis, they are covered up again during interphase until the next mitosis.

In view of the observation that synthesis of host DNA is required for this membrane change and that it occurs regularly at a certain point in the cell cycle, we would like to postulate as a working hypothesis that surface structural changes during mitosis are required for later events in the mammalian cell cycle (G1, S, G2, M) and that nuclear events are required for surface changes such as those involved in cell division. Using the surface changes described, we should be able to test this model experimentally.

Acknowledgements

This investigation was aided by U.S.P.H.S. Grants CA-10151 and 1-K4-CA-16,765, and by the Anita G. Mestres and Whitehall Foundations.

REFERENCES

ABERCROMBIE, M. and AMBROSE, E. J. (1962) *Cancer Res.* **22**, 526.

BEN-BASSAT, H., INBAR, M. and SACHS, L. (1970) *Virology* **40**, 854.

BENJAMIN, T. L. (1970) *Proc. natn. Acad. Sci. U.S.A.* **67**, 394.

BENJAMIN, T. L. and BURGER, M. M. (1970) *Proc. natn. Acad. Sci. U.S.A.* **67**, 929.

BIDDLE, F., CRONIN, A. P. and SANDERS, F. K. (1970) *Cytobios.* **5**, 9.

BURGER, M. M. (1968) *Nature, Lond.* **219**, 499.

BURGER, M. M. (1969) *Proc. natn. Acad. Sci. U.S.A.* **62**, 994.

BURGER, M. M. (1970) In *Permeability and Function of Biological Membranes*, ed. Bolis, L. Amsterdam: North Holland.

BURGER, M. M. and GOLDBERG, A. R. (1967) *Proc. natn. Acad. Sci. U.S.A.* **57**, 359.

COMAN, D. R. (1944) *Cancer Res.* **4**, 625.

CUNNINGHAM, D. D. and PARDEE, A. B. (1969) *Proc. natn. Acad. Sci. U.S.A.* **64**, 1049.

ECKHART, W., DULBECCO, R. and BURGER, M. M. (1971) *Proc. natn. Acad. Sci. U.S.A.* **68**, 283.

FOX, T. O., SHEPPARD, J. R. and BURGER, M. M. (1971) *Proc. natn. Acad. Sci. U.S.A.* **68**, 244.

HAKOMORI, S. and MURAKAMI, W. T. (1968) *Proc. natn. Acad. Sci. U.S.A.* **59**, 254.

HÄYRY, P. and DEFENDI, V. (1970) *Virology* **41**, 22.

HOLLEY, R. W. and KIERNAN, J. A. (1968) *Proc. natn. Acad. Sci. U.S.A.* **60**, 300.

INBAR, M., RABINOWITZ, Z. and SACHS, L. (1969) *Int. J. Cancer* **4**, 690.

INBAR, M. and SACHS, L. (1969a) *Proc. natn. Acad. Sci. U.S.A.* **63**, 1418.

INBAR, M. and SACHS, L. (1969b) *Nature, Lond.* **223,** 710.

JANSONS, V. K., SAKAMOTO, C. K. and BURGER, M. M. (1970) *Fedn Proc. Fedn Am. Socs exp. Biol.* **29,** 410.

JOHNSON, R. (1968) Senior Thesis, Princeton University.

PARDEE, A. B. (1964) *Natn. Cancer Inst. Monogr.* **14,** 7.

POLLACK, R. E. and BURGER, M. M. (1969) *Proc. natn. Acad. Sci. U.S.A.* **62,** 1074.

POLLACK, R. E., GREEN, H. and TODARO, G. J. (1968) *Proc. natn. Acad. Sci. U.S.A.* **60,** 126.

RUBIN, H. (1970) *Science* **167,** 1271.

SELA, B. A., LIS, H., SHARON, N. and SACHS, L. (1970) *J. Membrane Biol.* **3,** 267.

SHEPPARD, J. R., LEVINE, A. J. and BURGER, M. M. (1971) In preparation.

STOKER, M. and RUBIN, H. (1967) *Nature, Lond.* **215,** 171.

UHLENBRUCK, G., PARDOE, G. E. and BIRD, G.W. G. (1968) *Naturwissenschaften* **55,** 347.

WEISSMANN, G., TROLL,W., VAN DUUREN, B. L. and SESSA, G. (1968) *Biochem. Pharmac.* **17,** 2421.

DISCUSSION

Dulbecco: Can you characterize the cell surface sites yet, and especially whether the sites that you detect by different techniques are the same?

Burger: By the "same" we mean the following. If we isolate from the cell surface by a hypotonic shock (Burger 1968) material which has specific agglutination inhibitory properties, we obtain particles from transformed as well as normal cells with such properties. If we then isolate a glycoprotein from these particles by treatment with phenol, we find that this glycoprotein fraction has the agglutination inhibition capacity. A glycolipid fraction inhibits also but there seems to be much less of it. We feel therefore that it is primarily the glycoprotein which is the receptor. By electrophoresis we have so far found one band which shows this hapten inhibition; this band contains N-acetylglucosamine, as predicted from the hapten inhibition; we have, however, not yet shown that it contains di-N-acetyl-chitobiose, the disaccharide of N-acetylglucosamine which was a more powerful inhibitor than N-acetylglucosamine, or where in the glycoprotein it is located. One has to prove that there are two glycoproteins with the same sugar sequence in the material from transformed cells and from normal cells, before one can say the sites are identical.

Dulbecco: Of course the preparation from which you start may not be pure, and if you already have a mixture of sites, the heterogeneity may not be detected by the immunological technique.

Burger: Particularly since these lectins are not antibodies and have a wide range of specificity. This problem worries us more for the time being than the question of heterogeneity of the sites. Antibodies are much more specific than these lectins. Wheat germ agglutinin so far has specificity only for N-acetylglucosamine. This looks promising because so far no other sugar has been found to inhibit this reaction. In general a sugar may occur in α or β linkage, or of course in various places within the oligosaccharide chain, and these may be the specificities picked out by antibodies, whereas many lectins (plant agglutinins) may not be able to detect such fine differences and may thus lead us to the erroneous assumption that the receptor from normal and transformed cells is the same.

Macpherson: In your shift-down experiments with ts-3 polyomatransformed BHK21 cells did you study the effect of inhibitors of DNA synthesis to see if they prevented the reappearance of the agglutination site?

Burger: We have not done that.

Mitchison: Have you any measure of the relative affinity of the agglutinin for N-acetylglucosamine and for the cellular binding site?

Burger: We have measured the uptake of agglutinin by transformed and normal cells, but that is not a true equilibrium measurement. By calculating the concentration of sites present on normal and transformed cells we get 20 times more sites per surface area in transformed cells than in normal cells. That, by the way, is an interesting number, because the 3T3 cells require just about one hour to go through mitosis and the cycle of our 3T3 cells is around 20 hours. Therefore that correlates very nicely with the results on site opening during mitosis.

As to the binding of the pure agglutinin for N-acetylglucosamine, this is not very high, about 10^{-3} to 10^{-4} M. We cannot yet compare the affinity for the site on the cell surface with that for N-acetylglucosamine since we have to get the actual and exact number of sites per cell first.

Rubin: Is there a different carbohydrate inhibitor for each of the nine agglutinins?

Burger: I hope that they are different; I mean that we as well as others have found other agglutinins to interact with the transformed cell surface and that their specificity varies. When I talk about surface changes on transformed cell surfaces in general I do not mean that they are specific for the wheat germ agglutinin site; there are apparently a variety of carbohydrates which become exposed during transformation. First, N-acetylglucosamine which reacts with wheat germ agglutinin; then jack bean agglutinin or concanavalin A reacts with α-mannose derivatives; the soybean agglutinin was found to be inhibited by N-acetylgalactosamine, and phytohaemagglutinin also by N-acetylgalactosamine. So various sugars are involved. We cannot yet say whether it is the same receptor carrying these different sugars, but my guess is that a series of macromolecules is exposed during transformation, from recent results in our laboratory.

Rubin: It seems then that many groups are exposed at the cell surface in the malignant transformation, rather than a single specific new group being made. This suggests widespread alteration of the cell surface which could be produced by the leakage of enzymes such as cathepsins, lipases, etc.

Burger: If the change is due to enzymes acting at the cell surface! It could also occur in enzymes involved in the biosynthesis or turnover of cell membranes inside the cell, giving rise to the surface effects we are observing.

Stoker: Is there any evidence of grouping of the sites—any overlap that you could detect?

Burger: What we would like to measure is how much overlapping there is between the effects of concanavalin A, wheat germ agglutinin and the other agglutinins; in this way we may be able to establish a map of relative distances between these sites.

Stoker: They needn't necessarily reflect a very general change in the surface, as Dr Rubin suggests; they could equally reflect a local change which is important for the uptake of growth substances, because all sites happen to be close together.

Shodell: To get some idea of the significance of the exposure of these sites in normal cell growth and also the coordination of the exposure of these sites during the normal cell cycle, do you know whether arresting cells in metaphase with colchicine also maintains cells in a state of exposure to the agglutinin or whether the binding of agglutinin itself during mitosis can prevent entry into the G1 phase?

Burger: We have not studied the effect of colchicine yet. Wheat germ agglutinin has one nasty property, namely that it is counteracted by serum components like orosomucoid, which contains a lot of N-acetylglucosamine. We found earlier that if we agglutinated L1210 cells and injected them into the peritoneal cavity of a mouse, the animal survives for only the same length of time as the animals injected with untreated L1210 cells. We then injected the agglutinated L1210 cells into the mouse peritoneal cavity and immediately aspirated them back again, and that half minute in the peritoneal cavity was enough time to dissociate the cells. This suggested that peritoneal fluid might contain a material which would dissociate cells and we therefore tested serum and found that it also dissociated cells. This inhibitor is in the α-globulin fraction.

Pontecorvo: How many of these agglutinins have lymphocyte-stimulating activity? What sort of general connexion do you see between the events you have investigated and mitotic stimulation of lymphocytes?

Burger: Nobody has so far looked into the question of stimulation of lymphocytes to blast formation with the multitude of agglutinins known; it will have to be done, however. Apart from concanavalin, phytohaemagglutinin is the only one so far tested and found to be acting as a mitogen as well as reacting more with polyoma-transformed cells than with 3T3 cells (Burger, unpublished observation). Lymphocytes are not agglutinated with wheat germ agglutinin and so the sites on lymphocytes are apparently not exposed or not present at all. Incidentally, erythrocytes are agglutinated by wheat germ agglutinin.

Weiss: Aub, Sanford and Cole (1965) showed that several kinds of wandering cells are agglutinated by this agglutinin.

Cunningham: You stated in your abstract that it is possible to reverse the growth behaviour of transformed cells under crowded conditions to that of normal cells by covering up the exposed binding sites. Could you tell us more about this?

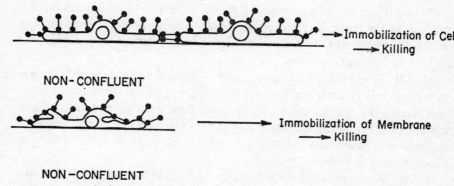

CONFLUENT

→ Immobilization of Cell
→ Killing

NON-CONFLUENT

→ Immobilization of Membrane
→ Killing

NON-CONFLUENT

→ Regular Growth Not Inhibited

Fig. 1 (Burger). A model for the differential effects of divalent agglutinin (killing) and monovalent agglutinin (no killing) which led us to the use of monovalent agglutinin for studies on growth control.

Burger: We have investigated whether wheat germ agglutinin blocks surface sites in transformed cells and stops overgrowth. That didn't work for the reason just given, namely that large concentrations of serum remove the wheat germ agglutinin. So some time ago we did some work with concanavalin A and showed that it kills transformed cells and normal cells, and Dr Sachs finds the same (Shoham, Inbar and Sachs 1970). We concluded that this was because concanavalin, having two sites, binds cells together as they crawl over each other and immobilizes them; or alternatively, in a non-confluent layer, it may bind two sites which are close together on the surface membrane and stiffen the cell surface, preventing growth and survival, if flexibility is important for these (Fig. 1).

We therefore attempted to cleave concanavalin A to make monovalent agglutinin, by treating the divalent material with chymotrypsin (Burger and Noonan 1970) or trypsin. (Steinberg and Gepner, in preparation)

The split agglutinin no longer agglutinated our test cells but inhibited agglutination by the uncleaved molecule. If we added monovalent concanavalin A to polyoma-transformed 3T3 cells, with increasing doses we could bring down the final cell number to a saturation density which was essentially the same as that of normal 3T3 cells (see Fig. 2). We ruled out a toxic effect on the cells by seeding different concentrations of transformed cells and adding the same suboptimal amount of monovalent wheat germ

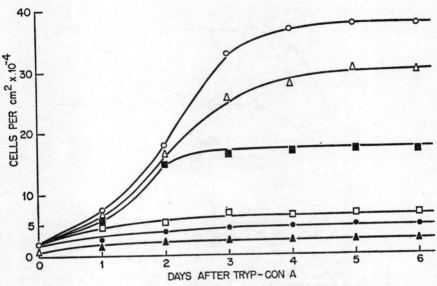

Fig. 2 (Burger). Effect of adding monovalent agglutinin to polyoma-transformed 3T3 cells.
o, Py3T3 cells.
△, Py3T3 cells + 10 μg/ml monovalent agglutinin.
■, Py3T3 cells + 25 μg ml monovalent agglutinin.
□, Py3T3 cells + 50 μg/ml monovalent agglutinin.
●, Py3T3 cells + 75 μg/ml monovalent agglutinin.
▲, 3T3 cells.

agglutinin; all the cells grew to the same density, which was four or five times less than the saturation density of the transformed cells. So growth was being blocked.

Mr K. Noonan and I ruled out an irreversibly toxic effect of the monovalent agglutinin in another experiment. Polyoma-transformed 3T3 cells were grown to confluency in the presence of monovalent concanavalin A and then hapten-inhibitor, which we knew to disrupt the specific cell surface–agglutinin interaction, was added (Fig. 3). The transformed cells continued to grow again at this point until they reached the same density as the untreated transformed cells (Burger and Noonan 1970). This may be evidence that we have saturated certain surface sites which are important in overcoming density dependent inhibition of growth. Normal cells seem to have some sort of covering layer which is involved in the density dependent inhibition of growth. Transformed cells lack this layer or it is misplaced, which is presumably why they pile up. This may happen in conjunction with or independently of sensitivity changes to serum factors.

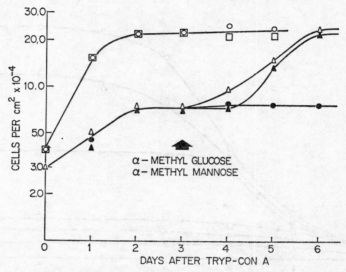

FIG. 3 (Burger). Effect of adding monovalent agglutinin and hapten inhibitor to polyoma-transformed 3T3 cells.
○, Py3T3 cells.
●, Py3T3 cells+50 μg/ml monovalent agglutinin.
□, Py3T3 cells+50 μg/ml monovalent agglutinin+10^{-2} M-α-methylglucoside for 12 hours.
△, Py3T3 cells+50 μg/ml monovalent agglutinin+10^{-2} M-α-methylglucoside or α-methylmannoside added at the third day.
▲, Py3T3 cells+50 μg/ml monovalent agglutinin+10^{-3} M-α-methylglucoside added at the third day.

If one adds an artificial cover layer, one can get back the inhibition of growth, characteristic of the normal cells. This artificial cover layer of monovalent agglutinin is obviously not an exact replacement of the natural cover layer. Thus apparently the nature of the surface site cover may not be of critical importance, but the sites covered up in normal cells are specific sites for transformed cells and whether they or nearby groups are covered or exposed may be the decisive factor for growth control.

Weiss: On one hand, agglutination sites appear to be exposed on transformed cells; on the other hand, Dr Montagnier was showing us that transformed cells have a thick coat of mucopolysaccharide and Bernhard has suggested that this coat shields cells from contact inhibition (Martinez-Palomo, Brailovsky and Bernhard 1969). Is this a real paradox or a false one?

Burger: This ruthenium binding layer could simply be due to over-synthesis of the material containing the surface receptor which is covered up in the normal cell.

Montagnier: The increased binding of ruthenium red does not mean necessarily that there are thicker layers of stainable material at the cell surface. It only suggests that there are more negative sites where the dye can precipitate.

REFERENCES

AUB, J. C., SANFORD, B. H. and COLE, M. N. (1965) *Proc. natn. Acad. Sci. U.S.A.* **54,** 396.

BURGER, M. M. (1968) *Nature, Lond.* **219,** 499–500.

BURGER, M. M. and NOONAN, K. D. (1970) *Nature, Lond.* **228,** 512–515.

MARTINEZ-PALOMO, A., BRAILOVSKY, C. and BERNHARD, W. (1969) *Cancer Res.* **29,** 925–937.

SHOHAM, J., INBAR, M. and SACHS, L. (1970) *Nature, Lond.* **227,** 1244–1246.

REGULATION OF CELL MULTIPLICATION IN TISSUE CULTURE

RENATO DULBECCO

The Armand Hammer Center for Cancer Biology, The Salk Institute for Biological Studies, San Diego, California

IN this paper I shall discuss models for the mechanism of the regulation of multiplication of animal cells which are suggested by the results of recent experiments, some obtained with uninfected cells, others after infection with oncogenic viruses (polyoma virus or SV40). Viruses allow the use of genetical methods which are a much needed tool in this field.

RESULTS WITH UNINFECTED CELLS

It is clear from a number of experimental results, some of them reported in this volume, that cell multiplication in tissue cultures depends on three conditions, besides the availability of precursors: the nature of the support for the cells; the density of the cell population; and the presence in the medium of serum and other substances. The role of the support is evident with many cell types which, in certain media, grow in contact with a glass or plastic surface but not in suspension. The role of cell density is shown by the slowing down of multiplication in crowded cultures. The requirement for serum factors has been discussed by Holley (pp. 3–10).

It has been uncertain for a long time how crowding operates: some investigators have taken it for granted that it results from inhibitory influences generated by cell-to-cell contact (so-called "contact inhibition of growth"), whereas others have proposed that it is produced by a reduced accessibility of the cells to serum. Two kinds of experimental evidence now tend to exclude the idea that the requirement for serum causes crowding inhibition. One type of evidence is based on the study of two parameters of cell growth in cultures: the *topoinhibition*, which is related to crowding inhibition, but is measured in the absence of serum; and the *serum requirement in a wound*, which is measured in the absence of crowding (Dulbecco 1970a). The two parameters are determined in a way that tends to exclude a reciprocal interference and therefore it becomes possible to study how

71

they are correlated with each other, using many different kinds of cells. The result of this comparison shows that the two parameters are uncorrelated: furthermore, epithelial cells show high topoinhibition but have no wound serum requirement. The other type of evidence was obtained using a viral mutant, and I shall return to that later.

The topoinhibition and wound serum requirement parameters are measured by studying the ability of cells to synthesize DNA. If other cell functions are studied, different serum requirements are detected, such as for survival, movement and mitosis, and for counteracting topoinhibition. Some of these requirements may be satisfied by different factors in the serum.

On the basis of these results it seems likely that topoinhibition is the consequence of cell-to-cell contacts, leaving it undecided whether or not it is related to contact inhibition of movement (Abercrombie and Heaysman 1954). The inhibitory effect of contacts may result from the activation of specific topoinhibition receptors on the cell surface upon contact with suitable topoinhibition-generating sites on the surface of other cells; if so, inhibition would only result from special types of contact between cells. Topoinhibition may be a fundamental signal for the regulation of cell multiplication, producing a negative control.

The negative influence of topoinhibition can be alleviated in three ways without changing the serum concentration: by transforming the cells with an oncogenic virus (Dulbecco 1970a); by a brief exposure to trypsin, as discussed by Dr Burger (pp. 45–63); and by wounding. The activation of DNA synthesis in cells that migrate into a wound in a confluent culture layer appears to result from loss of topoinhibition because it can be connected with loss of contact with other cells (Dulbecco and Stoker 1970).

When cells become insensitive to topoinhibition by transformation or after the action of trypsin, they also undergo surface changes revealed by the appearance of sites for a wheat germ agglutinin (WGA), as discussed by Dr Burger. It is not clear whether the WGA sites appearing in the various conditions are identical in their protein components: only when suitable purification is achieved will it be possible to answer this question.

There seems to be, therefore, a mutual exclusion between the presence of WGA sites and the hypothetical topoinhibition (TI) receptor on the surface of the same cell. The cell surface may, therefore, exist in two states: a *growing state* (characterized by the presence of WGA sites and the absence of TI receptors), or a *resting state*, with an opposite combination. Changes from one state to the other may occur either by some generalized change of the surface membrane (such as the loss of some layer or the profound reshuffling of layers) or by a rearrangement of a basic component (for example, subunits containing both WGA and TI receptors). An exception to the regular mutual exclusion is the line F1 101 (Pollack, Green and Todaro 1968), since, as shown by Dr Burger, it contains very few WGA sites but it is also insensitive to topoinhibition (Dulbecco 1970a). A possible explanation is that in the cells of this line the correlation between the two sites is broken by a mutation in a cell surface component.

RESULTS WITH POLYOMA VIRUS

These experiments were done using either the wild-type virus or temperature-sensitive mutants, which are impaired in multiplication at high temperature (39°C) but multiply well at low temperature (32°C) (Eckhart 1969). In one strain, the mutation is in gene 2 and at high temperature it prevents replication of the viral DNA but not stimulation of cellular DNA synthesis (Fried 1970). In the other strain, the mutation is in gene 5. The effects of the latter mutation are manifest in either transformation or lytic infection (Dulbecco and Eckhart 1970; Eckhart, Dulbecco and Burger 1971). BHK cells transformed by the gene 5 mutant have, like cells transformed by wild-type virus, low topoinhibition and high wheat germ agglutinating activity at low temperature, but at high temperature have the opposite properties. Thus, the surface of these cells is in the growing state at low temperature and in the resting state at high temperature. Since both properties are controlled by the same viral gene the mechanisms / by which they are formed must share a common step. The nature of this step could be different, depending on how the correlation between the two characters is maintained, as discussed above. The study of these transformed cells also shows that other characters are not correlated with topoinhibition or WG agglutination: they are the ability to grow in agar and the wound serum requirement, which are identical in cells transformed by the mutant or by wild-type virus, and in either case not temperature-dependent. This result supports the independence of topoinhibition from wound serum requirement, as indicated above.

Balb/3T3 cells lytically infected by the gene 5 mutant are temperature-sensitive both in the activation of cellular DNA synthesis and in the appearance of the WGA sites, which again demonstrates a common step in the generation of these two changes by the virus. Both changes occur and are temperature-insensitive after infection with the viral strain mutated in gene 2, showing that they do not require viral DNA synthesis. However, the WGA site does not appear if DNA synthesis is blocked (reversibly) by hydroxyurea (Eckhart, Dulbecco and Burger 1971). The inference is that the appearance of the WGA site requires cellular DNA synthesis. Some polyoma mutants (Benjamin and Burger 1970) stimulate cellular DNA synthesis but not the appearance of the WGA site, and probably affect another step in the generation of the site.

The results with the mutant in gene 5 suggest that in the lytic infection the activation of DNA synthesis is the primary effect, and the surface changes (WGA site) secondary. Since the surface changes in turn can stimulate DNA synthesis, through decrease of topoinhibition, a positive feedback might arise, which would maintain normal cells in a growing state once this state is reached. Clearly, this is not the case, and the feedback, if it exists, must therefore be dampened. If, however, the feedback can last for several generations, progressively decreasing, it may give rise to the phenomenon known as "abortive transformation", usually attributed to the continued expression of viral genomes for several generations (Stoker 1968). Lack of expression of the T antigen in most 3T3 cells abortively transformed by SV40 (Todaro, Scher and Smith 1971) would seem to support the hypothesis of a positive feedback persisting after a brief expression of the viral genome, although other interpretations are possible. The hypothesis could be tested by halting cellular DNA synthesis for some time, since the feedback would be thus interrupted, and cells should soon resume a normal phenotype.

In stable transformation, the primary role of the virus could be to activate directly cellular DNA replication at each cycle; the surface changes could be secondary. The observation that the transformed cells contain new antigenic determinants in their surface is not incompatible with this mechanism of transformation. In fact, the new antigens could be cellular and normally masked; if viral, they could be specified by a viral gene different from gene 5; or they could be specified after cellular DNA synthesis has been activated. In the latter model, the viral gene product, which would finally end up in the surface membranes of the cells, would have to be membranotropic. This suggests a more detailed model for the control of cellular multiplication by the virus. The viral protein would first enter into the nuclear membrane, interacting with the DNA-synthesizing centres, and causing DNA synthesis to be initiated; later it would reach the plasma membrane.

The result of the hydroxyurea experiment tends to exclude that the primary change induced by the virus is in the cell surface and that activation of cellular DNA synthesis is secondary. Abolition of topoinhibition therefore may not be the main controlling mechanism. What then would be the meaning of the virus-induced changes in the surface membrane of the cells? It can be suggested that they are essential for the steps following DNA replication, such as the G2 phase and mitosis, since profound surface changes are probably required for the division of the cells. In fact, the serum requirement for mitosis is greatly changed after infection with polyoma virus (Dulbecco 1970b).

CONCLUSION

The following picture of cellular growth regulation emerges from the experiments discussed above. The cells would be able to receive and transmit to the nucleus both positive and negative external signals; their growth state would depend on the balance of these signals. In tissue culture, positive signals are provided by the substrate (glass or plastic); negative signals by intercellular contacts (topoinhibition). Macromolecular substances in the medium participate in various ways. Serum, for instance, counteracts topoinhibition and thus promotes growth by reducing the negative signals. Serum is also required for preventing the denaturation of cellular (surface?) components (Dulbecco 1970a). Some serum components may have a direct positive effect because serum allows the proliferation of certain untransformed cells in suspension in agar, where the normal positive signals of the substrate are absent. The negative signals of topoinhibition would normally overcome the positive signals. However, simple removal of topoinhibition would not be enough to elicit the initiation of DNA synthesis in the absence of positive signals, since most transformed cells do not grow in suspension without serum.

The positive balance of the signals reaching the cell surface would somehow be conveyed, probably by a chemical positive effector, to the nucleus, where it would activate centres on the nuclear membrane responsible for initiating DNA synthesis. This event, in turn, would cause (or allow) changes of the cell surface required for mitosis.

Infection with oncogenic viruses would either generate the positive effector or modify the centres on the nuclear membrane so as to activate DNA synthesis when the regular effector is present in small quantity. Trypsinization of the cells in confluent layers would reduce topoinhibition, perhaps by removing its receptors, and would allow the persistent positive signals of the substrate to come through. Serum may reduce topoinhibition by shielding its receptors. Migration of the cells into a wound would reduce topoinhibition by reducing intercellular contacts. An important corollary would be that a positive signal cannot be generated solely by a structural modification of the cell surface.

SUMMARY

This article examines the role of factors controlling the multiplication of cells *in vitro* (topoinhibition, serum) and the influence of infection by oncogenic DNA-containing viruses on cell growth. The effects of temperature-sensitive mutants of polyoma virus permit an analysis of the relationship between activation of cellular DNA synthesis, topoinhibition and the production of surface changes, revealed by a wheat germ agglutination. The results lead to a proposal on the regulatory organization of cells *in vitro*.

Acknowledgement

This work was supported by a National Cancer Institute Grant, No. CA-07592.

REFERENCES

ABERCROMBIE, M. and HEAYSMAN, J. E. M. (1954) *Expl Cell Res.* **6**, 293–306.
BENJAMIN, T. and BURGER, M. M. (1970) *Proc. natn. Acad. Sci. U.S.A.* **67**, 929.
DULBECCO, R. (1970a) *Nature, Lond.* **227**, 802–806.
DULBECCO, R. (1970b) *Proc. natn. Acad. Sci. U.S.A.* **67**, 1214–1220.
DULBECCO, R. and ECKHART, W. (1970) *Proc. natn. Acad. Sci. U.S.A.* **67**, 1775–1781.
DULBECCO, R. and STOKER, M. G. P. (1970) *Proc. natn. Acad. Sci. U.S.A.* **66**, 204–210.
ECKHART, W. (1969) *Virology* **38**, 120–125.
ECKHART, W., DULBECCO, R. and BURGER, M. M. (1971) *Proc. natn. Acad. Sci. U.S.A.* **68**, 283.
FRIED, M. (1970) *Virology* **40**, 605–617.
POLLACK, R. E., GREEN, H. and TODARO, G. J. (1968) *Proc. natn. Acad. Sci. U.S.A.* **60**, 126–133.
STOKER, M. (1968) *Nature, Lond.* **218**, 234–238.
TODARO, G. J., SCHER, C. D. and SMITH, H. S. (1971) This volume, pp. 151–162.

DISCUSSION

Montagnier: Dr Dulbecco, do the arguments you have presented rule out the possibility that the hypothetical change at the nuclear membrane is not the cause, but a consequence of an earlier change at the cell surface? There may be a functional connexion between the two membranes, such as a dependency on common products, so that an early event at the cell surface may result in a later change on the nuclear membrane, and conversely. Experiments with hydroxyurea cannot rule out the possibility that in some situations the first event comes from the cellular surface rather than from the nuclear membrane.

Dulbecco: I agree, but the problem is how to attack the question of whether one or the other cell component has a role. I don't see any way to do this except by accumulating mutants.

Rubin: How much of your thinking depends on the interpretation of the effects of hydroxyurea? All the cytostatic agents that have been examined for this property (which included excess thymidine, colchicine and actinomycin) have been found greatly to increase the amount of free hydrolytic enzymes in the cell, including ribonuclease and DNase, first by Erbe and co-workers (1966) and more recently by Lambert and Studzinski (1969). So one has to be concerned about interpreting the effects of inhibitors of DNA, which may be doing many other things, including having effects even directly on the surface.

Dulbecco: There is good evidence that this doesn't happen, because the effect of hydroxyurea is completely reversible. If you leave polyoma-infected cells for 30 to 40 hours in hydroxyurea the sites are not made; when you remove the hydroxyurea the sites appear with a lag that is similar to the lag following infection in the absence of hydroxyurea, or even shorter. These cells continue to grow perfectly well.

Rubin: This is also true of the thymidine block. The cells recover completely, even though the level of nuclease may be increased five-fold. Some of the enzymes remain at high levels even after the cells start dividing and this persists for several generations. So the fact that cells grow after an inhibitor is removed does not argue for the specificity of the inhibitor.

Bürk: If you give a 12-hour pulse of 10 per cent serum followed by 12 hours of 2 mM-hydroxyurea and then remove the hydroxyurea, you get only one cell cycle. This suggests that the effect of hydroxyurea does not persist. Your prediction is that cells would carry on for some time?

Rubin: No. The cells return to the mitotic cycle after a high-thymidine block, but the inhibitor could be affecting other processes in the cell besides division, which may not be the process most sensitive to these agents. Lambert and Studzinski were looking at different kinds of enzymes. Some of them are unaffected and the cells go through a normal synchronous cycle; some of them alter, particularly alkaline DNase, which goes up about five-fold and remains up for several cell generations which take place normally. What this could do to a cell, I don't know; it's not stopping its growth. One has to take these undesirable events into consideration in interpreting the phenomenon that one is investigating.

Taylor-Papadimitriou: Dr Dulbecco, you suggested that one could test the effect of hydroxyurea on abortive transformation, but how would one be able to measure this, since all the manifestations of abortive transformation, except the appearance of the site, depend in some way on DNA synthesis? Perhaps one could follow the time course of the induction of DNA synthesis and the appearance of the site. We have data on the time at which DNA synthesis begins in serum-depleted BHK21 cells infected with polyoma virus (Taylor-Papadimitriou, Stoker and Riddle 1971). If one had comparable data for the appearance of the site, one might be able to detect which event occurred first in this system.

Dulbecco: Nevertheless the same approach cannot be used to establish the relationship between the appearance of the sites and the induction of DNA synthesis in acute infection, because they appear gradually over roughly the same time period; only by using an inhibitor can you clearly see the interrelation.

Stoker: The change in cell arrangement is certainly more or less coincident with the onset of DNA synthesis, which, if it reflects sites, would give the same difficulty.

Montagnier: Meyer, Birg and Bonneau (1969) have studied the kinetics of the appearance of the cell surface antigen in BHK cells infected with polyoma virus. The antigen appears quite early in the viral cycle, at about 15 hours after infection, and therefore before the induction of DNA synthesis.

Stoker: In the transformed cells the virus might be inserting a protein with affinity for membranes which affected DNA synthesis, not just at one site, but more generally in the nuclear membrane. Is there any evidence of any binding of immunofluorescent wheat germ agglutinin to the nuclear membrane of transformed cells?

Burger: Even though we have partially fixed normal mitotic cells to observe the mitotic pictures we have not done this on transformed cells with prefixation, because this raises the problem of artifacts, so I can't answer the question yet. We have however examined subcellular fractions of normal and transformed cells, to answer the question whether they aggregate. Nuclei from 3T3 cells aggregate and so do microsomal fractions; mitochondria don't aggregate very much. The problem in all these cases is once you open up the cell, you release lysosomes and lysosomal enzymes, small amounts of which open up sites.

Dulbecco: I. Lieberman looked at isolated nuclei in regenerating and normal liver by electrophoresis and he claimed to find marked electrophoretic changes in the nuclear membrane in the regenerating cells.

Warren: To my knowledge there are at least three lines of biochemical work related to cell surfaces where differences between normal and transformed cells have been found. These differences change with the cell cycle. For instance, Dr Burger has shown that something disappears in transformed cells so that sites are revealed; apparently normal cells have such sites but they are covered, and perhaps the sites in these two types of cells are the same. The sites are revealed on the normal cell surface in mitosis, so that one might compare not just normal and transformed cells but, rather, normal cells in mitosis with transformed cells. To put it another way, the transformed cell may have a persistently mitotic character as far as its cell surface is concerned.

Besides the work of Hakomori and Murakami (1968) and of Mora and Brady and their colleagues (Mora *et al.* 1969), there is the work of Drs P. W. Robbins and Ian Macpherson (1971) on glycolipids, many of which undoubtedly occur in the cell surface. They found three glycolipids that appeared in increased concentration in normal cells (Nil2) in plateau phase after they had stopped dividing and had started to pile up, whereas in transformed cells at high cell density they were still present in low concentration. These glycolipids would seem to appear in cells when the cells stop dividing.

In the third case, Dr Clayton Buck, Dr M. C. Glick and I found a fucose-containing glycoprotein on the surface of transformed cells in the log phase of growth (Buck, Glick and Warren 1970). When these cells stopped growing or slowed up a great deal or were prevented from dividing with thymidine, very little of this material could be found in the cell surface. So this is something that is found in transformed cells in log phase although it is also present in small quantity in untransformed cells in log phase.

Can one relate all these changes in cell surfaces to shifts, or slight losses of control, so that slippages are occurring in the different phases of the cell cycle and materials which should be more or less confined to the mitotic phase persist when the cell is transformed, giving the cell a persistently mitotic cell surface character? To speculate, this might account for the decrease in adhesiveness of transformed cells and malignant cells, which have decreased adhesiveness like cells in mitosis. Both transformed cells and normal cells in mitosis have increased "leakiness" and perhaps there are other common characteristics. There may be, then, a slipping in control so that what should happen in one particular phase of the cell cycle starts overlapping into other phases, upsetting the cell only slightly but enough to result in serious consequences for the cell and for the host.

Dulbecco: You are saying that once you start activating DNA synthesis, you go through mitosis and the surface changes are all normal consequences of these events?

Warren: Yes, and after that things are not quite right, just slightly out of phase.

Pontecorvo: This recalls the classical "precocity" theory of meiosis of C. D. Darlington. According to this the two kinds of cell division, mitosis and meiosis, are unified by making the assumption that in meiosis nuclear division starts too early, before the chromosomes have divided into chromatids. Thus a change as substantial as that between mitosis and meiosis can be accounted for by a shift in the timing of a normal process.

Warren: There may not be any radical differences between normal and transformed cells; both have the same components but in different amounts. The transformed cell may have a little too much of a particular component at a particular time to allow for completely normal functioning.

Dulbecco: But how did it get there? Consider a transformed cell with its hundreds of genes and enzymes; we know this is done by a virus and that there are one or two viral genes at most which do the whole transformation. How do you get two viral genes to give a hundred different consequences? You suggest that one of the viral genes simply activates continuously the machinery of DNA synthesis and everything depends on that; the cells are now continuously activated, never enter a resting phase, and therefore the characteristic changes of that phase do not take place. But there is another possibility, that a viral protein goes into the membrane. You say that there is nothing new in a transformed cell, but there are in fact viral proteins in the cell (maybe two), and perhaps one goes into the membrane and causes all the other changes. There could be a chain of events, once you put a new protein into the membrane, because you change the lipid–protein interaction in the membrane and so the overall lipid composition may be immediately changed. All the other enzymes associated with the membrane are now found in a different environment, so they all change quantitatively. In fact, the changes described in membrane enzyme activity are quantitative. It seems to me that the real question is whether there is a viral protein in the membrane.

Stoker: This isn't such a long way from what Dr Warren was suggesting, if the whole series of changes occurs because there are a few new membrane components which alter the whole structure.

Weiss: To reinforce what Dr Warren was saying, cells in late mitosis are not very adhesive and also bind agglutinins, and some normal cells, such as granulocytes, are like this all through the cell cycle. These cells are wandering cells *in vivo:* their locomotion *in vitro* is not inhibited by contact with fibroblasts. It is obvious from ciné films of cells in mitosis and of many virally transformed cells that their membranes are unstable.

Burger: From cell biologists I gather that surface bubbling seen during mitosis does not occur in early prophase, but rather later, at just about the time when we also see the onset of site exposure that I mentioned.

Stoker: Dr Burger, can you comment on the question of whether there is a positive feedback from the cell surface? The results of your experiment in which growth restarted after you had treated confluent cells with trypsin, and also that with monovalent ConA, might provide the strongest evidence. Do your results in these two experiments make Dr Dulbecco's suggestion less likely, or not?

Burger: We have no proof yet that surface feedback regulates DNA synthesis. Our experiments with trypsin-treated concanavalin A have shown that transformed cells which are in contact are going back to the 3T3 growth pattern. I'm not able to say anything about how normal cells like 3T3 cells behave if they receive high doses of trypsinized concanavalin A. One should theoretically be able to cover the sites exposed during mitosis with high doses of trypsinized concanavalin A. We just have not done that experiment so far. Such feedback, as I have suggested in my talk for normal growing cells, under uncrowded conditions, is nevertheless a useful working hypothesis suggesting itself from our finding. For the time being I believe it will be better to consider control under crowded conditions separately to avoid confusion. Here site exposure seems to bring about loss of growth control while covering re-establishes growth control in transformed cells. You may assume therefore that some message enters from the surface and has something to do with the regulation of division, or DNA synthesis.

Dulbecco: The vital question is whether there is a *positive* signal going from the surface to the centre or only *negative* signals. Topoinhibition relies on negative signals, for example.

Stoker: The monovalent ConA activity might suggest a positive signal?

Burger: What we measure, namely cell numbers, takes three days after addition of the agent that sets the whole thing in motion. Even though mitotic figures and DNA synthesis of course come much earlier we will have to move up our observation parameters close to the initial effects on the membrane before we can start studying this question of negative or positive control. It does not take much imagination, but just some wild speculation, to come up with a theory explaining the trypsin and ConA effect using negative signal growth control.

Rubin: Chicken cells turn on as fast as they do with any kind of stimulus. DNA synthesis starts in about 6 hours, and mitosis within 10–12 hours. All the cells have divided by 24 hours; then the cell numbers double.

Burger: The 3T3 cells react on a different time scale to the virus; DNA synthesis occurs much later (after 80 hours).

Rubin: I have long been bothered by the fact that when Rous cells transform they look very much as cells do in the early stages of infection with cytocidal viruses, when the cells just start to round up. The possibility is not ruled out that the effect of Rous sarcoma virus differs only quantitatively from that of the cytocidal viruses. The difference could be the rather negative one that Rous sarcoma virus doesn't damage the cell enough to kill it, but just enough to make it leaky and less adhesive. Dr Burger has examined the lytic cycle of infection with polyoma virus; have you looked at the early stages of the lytic cycle with ordinary lytic viruses to see whether you get unmasking of the sites there? I am thinking of something like Newcastle virus disease—something completely ridiculous.

Burger: No, we have not yet tried that.

Rubin: There is a tendency to think about the cell membrane as a somewhat static structure, but the cell membrane's structure is dependent not only on the nature of the constituent lipid and protein subunits but on the internal and external milieu, which determine the nature and extent of the non-covalent association between the subunits. Thus a virus need not insert new subunits into a membrane to alter its structure and function; it could accomplish this by changing the internal milieu, or by causing hydrolytic enzymes to leak from the lysosomes. Vasiliev and co-workers (1970) found that a variety of agents—ribonuclease, digitonin, hyaluronidase—would stimulate growth in density-inhibited cells. It is possible that a variety of surface-active agents could alter the membrane in a relatively non-specific way—by reducing surface charge and proton concentration, for example—and thereby stimulate growth. Proteases in general accomplish this. It may therefore be a futile quest to look for highly specific serum proteins, especially when they work across very wide species barriers.

Bürk: To extend that to the problem of the virus, if we assume that the DNA virus is a parasite on the cell's DNA-synthesizing machinery, the problem the virus has is that the cell presumably has a rule which says that when the cell has made one copy of DNA it cannot make another until the cell has divided, otherwise there is chaos in the amount of DNA per cell. This is terrible from the virus's point of view because it can only get itself copied once per cell division. The virus may be able to sit around and wait until some accident happens to the cell, such as Dr Rubin referred to for his chick cells and Dr Vasiliev's cells. After all, the virus gets in by some sort of accident. The virus may be able to wait for cell DNA synthesis to start but it has to break the cell's one-copy rule, and maybe we should look for something that operates as a signal saying "mitosis complete".

Stoker: Are you thinking about the successful virus in a lytic infection, rather than in transformed cells?

Bürk: I'm being very teleological, but transformation would be a special case of lytic infection in which the one-DNA-copy rule is not broken because virus DNA has become cell DNA (integration). You could then say that transformation is the consequence of a small amount of virus telling the cell continuously "You have completed mitosis. You may start DNA synthesis when you are ready." For instance, something might be made in the cytoplasm which is necessary for DNA synthesis and cannot get into the nucleus except when the membrane disappears in mitosis.

Dulbecco: The mutants do show that there is a strict correlation between viral function in acute infection and in transformation.

Rubin: Renato, have you examined the transformants produced by the temperature-sensitive mutant of polyoma at intermediate temperatures? Martin finds that his temperature-sensitive mutant of Rous sarcoma virus (Martin 1970) produces fusiform foci at intermediate temperatures, which eventually round up into typical Rous foci (Martin, unpublished). The fusiform foci resemble foci produced by Temin's "morph*f*" mutants (Temin 1960). The latter were considered to be a specific morphological type but they now appear to represent an intermediate degree of transformation.

Dulbecco: We haven't looked at this.

Pontecorvo: It seems helpful to look at the transformed cell as one in which the basic difference is one of the timing of one normal process and all the rest is a consequence.

Dulbecco: It seems to me that the only thing needed to make a transformed cell is the signal which at every new division tells the cells that they don't have to wait for an environmental change, but to go ahead and divide again.

Pontecorvo: Do you need that? We are seeing that most of the properties of transformed cells turn out to be properties of a normal cell which are slightly out of phase.

Dulbecco: Then why would there be a difference between abortive transformation and stable transformation? If the only thing to do is to make an initial change and then the thing will proceed, there should be no abortive transformation.

Pontecorvo: The initial change has to be repeated in every cycle, I agree.

Warren: One could alternatively think of transformation as being the consequence of a process normally occurring at a certain time during the cell cycle that persists, that the cell can't shake off.

Dulbecco: This discussion has familiar undertones. Various individuals are showing their original tendencies, and I can always recognize the generalists and the specialists! The generalists always think that what happens is a reflexion of a very general phenomenon; the specialists want to find one cause. Working with viruses, I have to be a specialist because I know that there is one cause (or two): one (or two) genes, which we identify by their mutants.

Rubin: I also work with viruses and I'm a generalist.

Dulbecco: You are not concerned with mutants!

Rubin: But Steven Martin, in my laboratory, is, and he finds evidence that mutants which appear to produce a new type of transformed cell are in reality producing a partial transformation. Given enough time, the transformation progresses to the classical type. The mutation may be in a rate-controlling element rather than a structural one.

Dulbecco: You mentioned quantitatively different effects which occur at different temperatures after infection with viruses carrying temperature-sensitive mutations. Certainly, at intermediate temperatures between that allowing normal gene function and that blocking it completely, the gene is partially blocked and therefore you have quantitatively different effects. But this does not mean that the introduction of the gene into the cells does not cause an all-or-none effect.

Stoker: In fact, I should have thought Dr Rubin's remarks rather reinforced this. Dr Burger, have you looked to see whether there is any difference in the wheat germ agglutinin binding in the *transformed* cells in relation to mitosis? If you limit the amount of the agglutinin put on, can you see any relationship to the cell cycle? What proportion of transformed cells fluoresce when you stain them with the wheat germ agglutinin? This might support Dr Warren's point that there is an extension of the mitotic state of the membrane.

Burger: All of the cells of a stable transformed line fluoresce.

Stoker: If you reduce the amount of monovalent ConA do they all fluoresce equally?

Burger: We haven't done that. The only way to do that properly would be to use our tritium-labelled agglutinin. Since non-synchronized cultures of 3T3 cells seem to have one-twentieth as many wheat germ sites as transformed cells and since mitosis takes one hour of the 20-hour cell cycle our results then suggest that exposure during mitosis equals the permanent exposure in transformed cells. These preliminary results, as I have said before, will have to be repeated and extended to other cells. Nevertheless our working hypothesis at the present time is that an infected cell goes into mitosis and then gets "stuck" in that mitotic surface configuration.

Bergel: It might be useful to compare cells transformed by viruses with those transformed *in vitro* by carcinogenic chemicals or X-irradiation (see Sachs 1966; Chen and Heidelberger 1969). There one has three types of transformed cells; would one find differences in their responses if they are exposed to the experimental situations of Dr Dulbecco and Dr Burger, and could one draw some conclusions from such differences about what really happens on and in the cells?

Burger: If you just look at the concanavalin site and the wheat germ site, the answer is yes, so far; all these transformations lead to the same change with regard to agglutinability, in the chemically, the virally and the X-ray transformed cells. Detailed quantitative analyses however become very important here as well as analyses of the specificity of the sites. We have found, for instance, that different viruses may expose different agglutinin receptor sites. It would also be interesting to see, if we test all our other agglutinins, whether other receptor patterns emerge in response to chemicals and X-rays.

Bergel: So it seems that though the controlling genome is different in the three types (one possibly mixed with viral nucleic acid, the second with chemically altered and the third with distorted nuclear material), the stability or instability characteristics of the transformed cells are roughly the same. This amounts to encountering the same or similar problems whatever the transforming agent is, unless one allows for the possibility that the chemical and radiological transformations are based ultimately on the transforming action of unknown viruses.

Stoker: The difficulty with these other systems is that it hasn't been possible to investigate them during the initiation state, but maybe this will become possible.

I think that there is a difference in the S surface antigen sites detected by immunofluorescence (Häyry and Defendi 1970). As I understand it there is an antigen which is detectable by immunofluorescence and mixed agglutination, which is present on SV40-transformed cells, but occurs on normal and polyoma-transformed cells only after trypsin treatment.

Rubin: I think the S antigen is also present in polyoma cells, from talking to Defendi.

Burger: I'm sure there is an S antigen for polyoma-transformed cells which is different from the SV40 S antigen, but the question is, does it react in Defendi's test after trypsinization of the normal cell?

Weiss: Is Forssman antigen revealed by trypsinization?

Macpherson: No.

Mitchison: How is the S antigen known to be separate from the transplantation antigen?

Macpherson: Variants of hamster cells transformed by SV40 have been described by Tevethia and co-workers (1968) that have S antigen but no detectable transplantation antigen.

Pontecorvo: A large number of chemically induced tumours are known to be antigenically unique.

Burger: The problem there is that one selects for differences.

Rubin: To return to the general theme, I don't know that there is so much difference between the generalists and specialists here. I am not questioning the existence of mutants (shades of Hinshelwood) nor that they are manifested in an altered protein. I am questioning whether the mutant product is necessarily responsible specifically and directly for the transformation, and whether this product is inactivated by the temperature elevation. For example, let us say that viral infection alters the internal milieu of the cell, and this in turn causes a permeability change. The temperature-sensitive mutant's effect on permeability may be less, and the increased temperature might merely allow cellular compensation to occur. There is a precedent of sorts for this. Lwoff (1962) had a strain of polio virus whose development was blocked by high temperature. He concluded that the mutation modified a structural gene for the synthesis of a viral polymerase. Seven years later the observations were reinterpreted so that the mutant was now merely a slow-growing variant which was vulnerable to nucleases released from lysosomes by high temperature (Lwoff 1969). The temperature-sensitive mutation was one which affected the rate of viral development, rather than a viral product which could be inactivated by high temperature. The temperature elevation affected a cellular process and only indirectly the more slowly developing virus mutant.

Dulbecco: There is a profound difference, however, between transformation produced by a virus and that caused by mechanisms involving primarily cellular genes, as may be the case in chemically induced or radiation-induced transformation: operationally, the enormous difference arises because in the viral case, you are concerned with viral genes, and can obtain mutants which reveal what the action of these genes is. The mutants allow one to identify links in the chain you describe. In the other case the genetic tool is lacking.

Stoker: Unfortunately, to get at the other links, cell temperature mutants are going to be needed and these will be much more difficult to obtain.

So far in the meeting we have gone from serum factors to surface sites and we have not tied them together very much; in fact Dr Dulbecco has put them in separate categories! However, the same topics will be recurring later and some synthesis may then appear.

REFERENCES

BUCK, C. A., GLICK, M. C. and WARREN, L. (1970) *Biochemistry* **9**, 4567.

CHEN, T. T. and HEIDELBERGER, C. (1969) *Int. J. Cancer* **4**, 166.

ERBE, W., PREISS, J., SEIFERT, R. and HILZ, H. (1966) *Biochem. biophys. Res. Commun.* **23**, 392–397.

HAKOMORI, S and MURAKAMI, W. T. (1968) *Proc. natn. Acad. Sci. U.S.A.* **59**, 254.

HÄYRY, P. and DEFENDI, V. (1970) *Virology* **41**, 22.

LAMBERT, W. and STUDZINSKI, G (1969) *J. cell Physiol.* **73**, 261–266.

LWOFF, A. (1962) *Cold Spring Harb. Symp. quant. Biol.* **27**, 159–174.

LWOFF, A. (1969) *Bact. Rev.* **33**, 390–403.

MARTIN, S. (1970) *Nature, Lond.* **227**, 1021.

MEYER, G., BIRG, F. and BONNEAU, H. (1969) *C.r. hebd. Séanc. Acad. Sci., Paris, Sér. D* **268**, 2848–2849.

MORA, P. T., BRADY, R. O., BRADLEY, R. M. and McFARLAND, V. W. (1969) *Proc. natn. Acad. Sci. U.S.A.* **63**, 1290.

ROBBINS, P. W. and MACPHERSON, I. A. (1971) *Proc. R. Soc. B* **177**, 49–58.

SACHS, L. (1966) In *Molekulare Biologie des malignen Wachtums*, ed. Holzer, H. and Holldorf, A. W. Berlin, Heidelberg and New York: Springer.

TAYLOR-PAPADIMITRIOU, J., STOKER, M. G. P. and RIDDLE, P. (1971) *Int. J. Cancer* **7**, 269.

TEMIN, H. (1960) *Virology*, **10**, 182.

TEVETHIA, S. S., DIAMANDOPOULOS, G. T., RAPP, F. and ENDERS, J. F. (1968) *J. Immun.* **101**, 1192.

VASILIEV, Ju.M., GELFAND, I. M., GUELSTEIN, V. I. and FETISOVA, E. K. (1970) *J. cell. Physiol.* **75**, 305–314.

MOLECULAR EXCHANGE AND GROWTH CONTROL IN TISSUE CULTURE

JOHN D. PITTS

Department of Biochemistry, University of Glasgow

IN the living animal most cells interact, either directly or indirectly, with other cells, thus allowing coordinated control of growth and function, but most of this interaction and coordination is lost when the cells are taken from the animal and grown in tissue culture. However, in some cases at least, a form of direct cell–cell interaction occurs in tissue culture and in certain specialized instances such interaction can result in the coordinated growth of two cell types in a mixed culture.

Direct cell–cell interaction in tissue culture in the form of intercellular junctions of low electrical resistance (low-resistance junctions) has been described by Potter, Furshpan and Lennox (1966). Such junctions, like those previously described in organ culture (Loewenstein and Kanno 1964, 1966), permit the ready passage of ions, whereas elsewhere on the cell surface the cytoplasmic membrane is relatively resistant to such ion passage.

A chance observation in our laboratory some years ago led to the discovery of a cell–cell transfer phenomenon which we called "metabolic cooperation" (Subak-Sharpe, Bürk and Pitts 1966). Whether or not this phenomenon is related to the formation of low-resistance junctions between cells is not known.

Variant cells of the polyoma-transformed hamster fibroblast line PyY (cell lines are described in Table I) which lack inosinic pyrophosphorylase (PyY–IPP⁻ cells) cannot incorporate [³H]hypoxanthine, but radioauto-graphic studies showed that such PyY-IPP⁻ cells can incorporate [³H]hypoxanthine when growing in contact with wild-type PyY cells (Subak-Sharpe, Bürk and Pitts 1969). Apparently contact is required, since PyY-IPP⁻ growing close to but not in contact with wild-type cells do not incorporate [³H]hypoxanthine. Something passes between the cells in contact, but what is it? In a previous publication we listed the various possibilities (Subak-Sharpe, Bürk and Pitts 1969) and these fall into two distinct classes:

Class 1: a ^{3}H-labelled compound is transferred (e.g. [3H]IMP or some other derivative of hypoxanthine which can be incorporated by IPP⁻ cells).

Class 2: something is transferred which endows the IPP⁻ cell with the ability to incorporate [3H]hypoxanthine (for example, the enzyme inosinic pyrophosphorylase, or nucleic acid containing information for the synthesis of the enzyme, or control molecules—the mutation in IPP⁻ cells has not been shown to be in the structural gene for the enzyme).

Further work showed that reciprocal exchange between two cell types can take place, and that the phenomenon is not restricted to IPP⁻ cells. PyY-APP⁻ cells (which lack the enzyme adenylic pyrophosphorylase and cannot incorporate [3H]adenine) were grown in confluent culture with equal numbers of PyY-IPP⁻ cells and duplicate cultures were exposed to either [3H]hypoxanthine or [3H]adenine. This clearly showed that contact between IPP⁻ cells and APP⁻ cells in the confluent culture resulted in phenotypic modification of the two cell types, [3H]hypoxanthine and [3H]adenine being incorporated by all the cells (IPP⁻ and APP⁻) in the mixed cultures.

This work has now been extended and the results will be described here, but not the experimental procedures which are described elsewhere (Bürk, Pitts and Subak-Sharpe 1968; Pitts 1971). The enzyme-deficient strains of the hamster fibroblast line BHK21/13 and of the mouse fibroblast line L929 were kindly provided by Dr John W. Littlefield (see Table I).

TABLE I

PARENTAL WILD-TYPE AND VARIANT CELL STRAINS USED

Parental wild-type	Variant	Enzyme deficiency
BHK (hamster fibroblast cell line BHK21/13[1])	BHK-IPP⁻ (T6A[2])	Inosinic pyrophosphorylase (E.C. 2.3.2.8.)
BHK	BHK-TK⁻ (B1[2])	Thymidine kinase (E.C. 2.7.1.21.)
PyY (polyoma-transformed BHK21/13 cell line[3])	PyY-IPP⁻ (TG1[4])	Inosinic pyrophosphorylase
PyY	PyY-APP⁻ ([5])	Adenylic pyrophosphorylase (E.C. 2.4.2.7.)
L (mouse fibroblast cell line L929[6])	L-IPP⁻ (A9[7])	Inosinic pyrophosphorylase
L	L-TK⁻ (B82[7])	Thymidine kinase

References:
1. Macpherson and Stoker (1962)
2. Marin and Littlefield (1968)
3. Stoker and Macpherson (1964)
4. Subak-Sharpe (1965)
5. Bürk, Subak-Sharpe and Hay (1966)
6. Sanford, Earle and Likely (1948)
7. Littlefield (1966)

Wild-type BHK cells incorporate [³H]hypoxanthine into acid-insoluble material and the extent of incorporation can be assessed by radioautography, if the numbers of silver grains found over a large number of individual cells are counted. The results of such a grain count are shown in Fig. 1a. Incorporation of [³H]hypoxanthine into wild-type cells results in a unimodal distribution of grain counts with a mean of 41 grains per cell. BHK-IPP⁻ cells grown separately in the presence of [³H]hypoxanthine and treated in the same way show a markedly different grain count distribution, with mostly only 0, 1 or 2 grains per cell (Fig. 1a). However, when equal numbers of wild-type BHK cells and BHK-IPP⁻ cells are grown together in confluent culture and again treated the same way, cell–cell interaction resulting in phenotypic modification of the IPP⁻ cells is clearly evident. The IPP⁻ cells which showed a grain count of only 0, 1 or 2 grains per cell when grown alone are completely masked in the mixed population, where no cells incorporate at such a low level.

The proportion of IPP⁻ cells in unlabelled duplicate mixed cultures was estimated by subculturing 1 per cent of the cells into medium containing [³H]hypoxanthine. After 8 hours the cells were examined by radioautography and single cells, not in contact with other cells, were scored as IPP⁻ (<3 grains per cell) or wild-type (many grains per cell). The proportion of IPP⁻ cells in different experiments varied from 39 to 53 per cent, showing that the absence of cells with a low grain count in the mixture illustrated in Fig. 1a is not a result of the selective death of 1PP⁻ cells.

An exactly analogous experiment was carried out with wild-type BHK cells and BHK-TK⁻ cells (which lack the enzyme thymidine kinase and cannot incorporate [³H]thymidine). Again the radioautographic technique distinguishes the TK⁻ cells from the wild-type cells (Fig. 1b), but the grain count distribution for the wild-type cells is confused by the presence of a small proportion of cells which do not incorporate [³H]thymidine, not because they lack thymidine kinase, but because they do not synthesize DNA (that is, pass through S phase) in the course of the experiment. However, it is clear that at least a large proportion (and probably all) of the BHK-TK⁻ cells incorporate thymidine because of cell–cell interactions, when grown in confluent culture with the wild-type BHK cells.

The nucleotide TMP (thymidine 5′-phosphate) is incorporated only very slowly (at less than 10 per cent the rate of incorporation of thymidine) into wild-type BHK cells, and the incorporation which is observed may not represent transport of TMP across the cytoplasmic membrane, but incorporation of degradation products of TMP formed in the medium. It has also been shown (Subak-Sharpe 1969) that AMP and GMP cannot enter wild-type PyY cells without loss of the phosphate group. We must

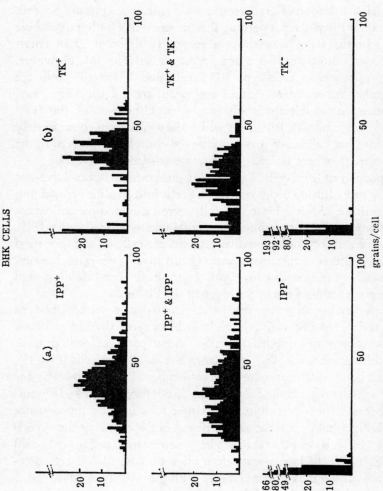

Fig. 1. Incorporation of [³H]hypoxanthine and [³H]thymidine into wild-type BHK cells, variant BHK cells and 1:1 cell mixtures. Confluent cultures of wild-type BHK, BHK-IPP⁻ and BHK-TK⁻ cells growing separately or together, as shown, were exposed for 12 hours to [³H]hypoxanthine (a) or [³H]thymidine (b). The cells were fixed and washed and radioautographs were prepared from samples of each culture; the silver grains located over >400 individual cells were counted.

conclude therefore that if transfer of nucleotides is the basis of this cell–cell exchange phenomenon, then the junctions which allow exchange must have different properties from the normal cytoplasmic membrane (cf. the low-resistance junctions described earlier). Furthermore, it has been shown that the acid-soluble pools of thymidine nucleotides are located predominantly in the cell nucleus (Adams 1969). Transport of such materials, therefore, from cell to cell (unless they travel by internuclear bridges, as suggested by Bendich, Vizoso and Harris 1967), requires rapid equilibration between the nuclear and cytoplasmic pools in both cells and also between the two cytoplasmic pools.

We next turned our attention to the mouse fibroblast line L929, and much to our surprise we found (as did Miss Chantal Favre, personal communication) that these cells do not interact in tissue culture in the way BHK and PyY cells do. When L-IPP⁻ or L-TK⁻ cells are grown in confluent mixed culture with wild-type L cells the variant phenotype is unaltered (Fig. 2a and b). Furthermore, this inability of L cells to interact with each other is dominant in mixed cultures of L-IPP⁻ cells with wild-type BHK cells or of BHK-IPP⁻ cells with wild-type L cells.

In retrospect, this result should not have been surprising, because L-IPP⁻ cells can be selected in dense culture by resistance to 6-thioguanine, whereas BHK-IPP⁻ must be selected in sparse culture, since contact of a variant cell with a wild-type cell in the presence of 6-thioguanine leads to the incorporation of the drug into both cells (Fujimoto, Subak-Sharpe and Seegmiller 1971). Because of the lack of cell–cell interaction this does not apply to L cells.

Mouse primary fibroblasts interact with PyY-IPP⁻ cells and BHK-IPP⁻ cells (Stoker 1967 and unpublished results), so why don't L cells, which are also mouse fibroblasts, interact? Has the strain lost the ability to form some essential membrane component or do L cells contain a controlling factor which prevents cell–cell interaction? Experiments with L–BHK hybrid cells induced with inactivated sendai virus (Harris and Watkins 1965) are in progress in an attempt to understand this problem.

Some attempt has also been made to discover what type of molecule is transferred in these interactions. Equal numbers of BHK-IPP⁻ cells and wild-type BHK cells were grown together in confluent culture (in unlabelled medium) for 18 hours to allow exchange to take place. At that time, they were rapidly subcultured into medium containing [³H]hypoxanthine, 99 per cent of the cells into one culture and 1 per cent into another; after incubation for a further 8 hours the cells were examined by radioautography. The confluent culture formed from 99 per cent of the original cells contained only incorporating cells (Fig. 3), but single cells examined in

5*

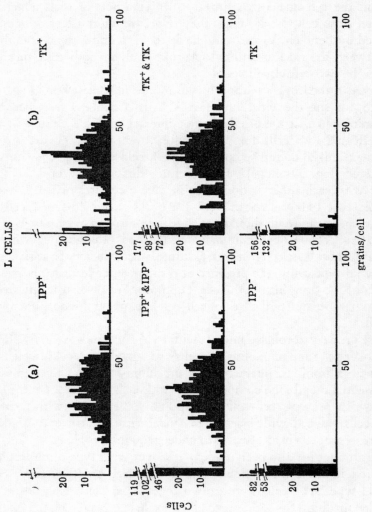

FIG. 2. Incorporation of [³H]hypoxanthine and [³H]thymidine into wild-type L cells, variant L cells and 1:1 cell mixtures. For experimental procedure see legend to Fig. 1.

the sparse culture formed from 1 per cent of the original cells were of two types: those with the incorporation characteristics of the BHK-IPP⁻ cells and those with the characteristics of the wild-type BHK cells (Fig. 3).

Reconstruction experiments using prolonged radioautographic exposures further showed that if the BHK-IPP⁻ cells in the sparse culture had incorporated [³H]hypoxanthine at the wild-type rate for only 10 minutes, this could have been detected.

BHK CELLS

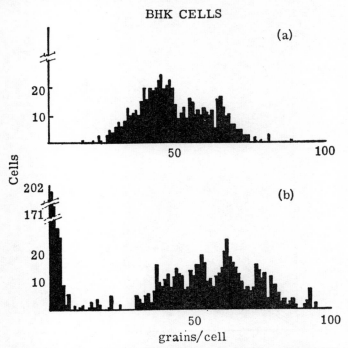

FIG. 3. Instability of contact-induced wild-type phenotype of BHK-IPP⁻ cells after separation from wild-type BHK cells. A confluent culture of mixed (1:1) wild-type BHK and BHK-IPP⁻ cells was subcultured into medium containing [³H]hypoxanthine, 99 per cent of the cells into one subculture (a) and 1 per cent into another (b). After 8 hours the cells were fixed and washed. Radioautographs were prepared from samples of each culture and the silver grains located over > 400 individual cells were counted. (In (b), only single cells not in contact with other cells were counted.)

The simple conclusion from this experiment is that the ability to incorporate [³H]hypoxanthine (that is, the enzyme itself or information for its synthesis; see p. 90) is not transferred from cell to cell, but instead, IMP or some other derivative of hypoxanthine which can be incorporated by IPP⁻ cells. However, if the enzyme were very unstable (with a half-life of only a few minutes) a similar result would have been obtained.

Further experiments were suggested by the selection techniques used to isolate cell hybrids from the mixtures containing one IPP⁻ and one TK⁻ cell type. Aminopterin blocks the *de novo* synthesis of IMP and TMP, but wild-type cells can grow in medium containing aminopterin if hypoxanthine and thymidine are added (HAT medium). However, because of their respective enzyme deficiencies, IPP⁻ and TK⁻ cells are unable to grow in HAT medium but IPP⁻–TK⁻ hybrids can grow.

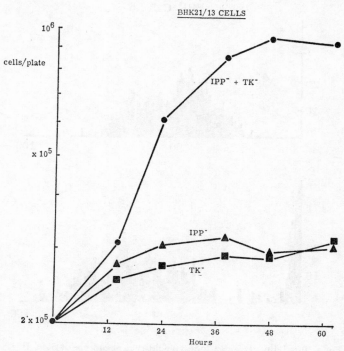

FIG. 4. Growth of BHK-IPP⁻ and BHK-TK⁻ cells separately and in mixed (1:1) culture, in medium containing hypoxanthine, aminopterin and thymidine (HAT medium).

We predicted, however, that cell–cell interactions should allow a dense culture of equal numbers of BHK-IPP⁻ and BHK-TK⁻ cells to grow, and, as shown in Fig. 4, this is so. The BHK-IPP⁻ cells and BHK-TK⁻ cells when cultured separately grow only to a small extent in HAT medium, but the mixed culture grows well and the doubling rate (at the maximum, between 12 and 24 hours) is similar to the doubling rate of wild-type BHK cells in the same HAT medium, indicating that the exchange process is very efficient. A similar experiment with L cell variants is illustrated in Fig. 5. The growth of the mixed culture (containing equal numbers of

L-IPP⁻ and L-TK⁻ cells) is inhibited to the same extent as the growth of the individual variant cells, again showing a lack of interaction between L cells.

The growth of the BHK-IPP⁻/BHK-TK⁻ cell mixture depends on a symbiotic relationship. It also depends on the pooling of genetic ability in the mixed culture where all cells show the same phenotype despite the

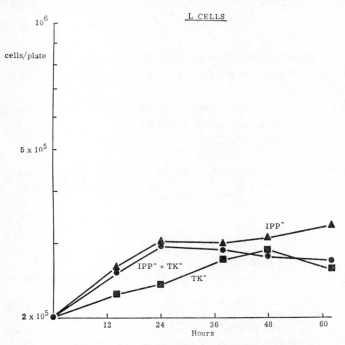

FIG. 5. Growth of L-IPP⁻ and L-TK⁻ cells separately and in mixed (1 : 1) culture, in HAT medium.

presence of two distinct genotypes. Growth is also dependent on the cell density and cells in mixed sparse cultures appear to grow only from groups of cells containing both cell types. Preliminary experiments also show that a 1 : 20 mixture of BHK-IPP⁻ and BHK-TK⁻ cells grows only slowly in sparse culture, but at a steadily increasing rate, and that the final confluent culture contains approximately equal numbers of the two variants; that is, the proportion of the two cell types in the mixed population is self-stabilizing.

This represents a form of growth control in tissue culture which depends on interaction and exchange between cells in contact, and the symbiotic nature of the system necessarily regulates the final proportions of the different cell types.

SUMMARY

1. Variant, enzyme-deficient strains of BHK cells, unable to incorporate hypoxanthine, adenine or thymidine, are phenotypically changed and incorporate these compounds when growing in contact with wild-type BHK cells.

2. This molecular exchange, which occurs between BHK cells in contact, does not occur between L cells in contact, or between L cells in contact with BHK cells.

3. Exchange between variant BHK cells allows a mixed population of two cell types to grow in selective medium which prevents the growth of either cell type alone.

4. The implications and possible mechanisms of molecular exchange between cells in tissue culture are discussed.

Acknowledgements

I am grateful to Miss Jacqueline Ferguson for able technical assistance and to Professor J. N. Davidson, F.R.S. and Professor R. M. S. Smellie who provided the facilities for carrying out this work, with the aid of grants from the Wellcome Trust and the Cancer Research Campaign.

REFERENCES

ADAMS, R. L. P. (1969) *Expl Cell Res.* **56**, 49–54.

BENDICH, A., VIZOSO, A. D. and HARRIS, R. G. (1967) *Proc. natn. Acad. Sci. U.S.A.* **57**, 1029–1035.

BÜRK, R. R., PITTS, J. D. and SUBAK-SHARPE, J. H. (1968) *Expl Cell Res.* **53**, 297–301.

BÜRK, R. R., SUBAK-SHARPE, J. H. and HAY, J. (1966). *Heredity* **21**, 343.

FUJIMOTO, W. Y., SUBAK-SHARPE, J. H. and SEEGMILLER, J. E. (1971) In preparation.

HARRIS, H. and WATKINS, J. F. (1965) *Nature, Lond.* **205**, 640–646.

LITTLEFIELD, J. W. (1966) *Expl Cell Res.* **41**, 190–196.

LOEWENSTEIN, W. R. and KANNO, Y. (1964) *J. Cell Biol.* **22**, 565–586.

LOEWENSTEIN, W. R. and KANNO, Y. (1966) *Nature, Lond.* **209**, 1248–1249.

MACPHERSON, I. A. and STOKER, M. G. P. (1962) *Virology* **16**, 147–151.

MARIN, G. and LITTLEFIELD, J. W. (1968) *J. Virol.* **2**, 69–77.

PITTS, J. D. (1971) In preparation.

POTTER, D. D., FURSHPAN, E. J. and LENNOX, E. S. (1966) *Proc. natn. Acad. Sci. U.S.A.* **55**, 328–336.

SANFORD, K. K., EARLE, W. R. and LIKELY, G. D. (1948) *J. natn. Cancer Inst.* **9**, 229–246.

STOKER, M. G. P. (1967) *J. Cell Sci.* **2**, 293–304.

STOKER, M. G. P. and MACPHERSON, I. A. (1964) *Nature, Lond.* **203**, 1355–1357.

SUBAK-SHARPE, H. (1965) *Expl Cell Res.* **38**, 106–119.

SUBAK-SHARPE, J. H. (1969) *Ciba Fdn Symp. Homeostatic Regulators*, pp. 276–288. London: Churchill.

SUBAK-SHARPE, H., BÜRK, R. R. and PITTS, J. D. (1966) *Heredity* **21**, 342–343.

SUBAK-SHARPE, H., BÜRK, R. R. and PITTS, J. D. (1969) *J. Cell Sci.* **4**, 353–367.

DISCUSSION

Abercrombie: Do L cells show a density dependence of growth which can't be ascribed to depletion of the medium?

Pitts: I have not measured the topoinhibition of L cells, but at high cell densities they begin to grow in suspension above the monolayer, so probably topoinhibition will be negligible.

Rubin: Have you tried normal mouse cells?

Pitts: We have tested primary mouse embryo fibroblasts and so has Michael Stoker in a similar system, and these do interact with BHK mutants, whereas wild-type L cells do not. There is a difference here between primary mouse fibroblasts and L mouse fibroblasts.

Weiss: What about transformed BHK cells—were they subject to topoinhibition by normal mouse cells?

Stoker: Yes; that was why the experiment was done.

One point arising from these old data may be relevant. When transfer was observed from wild-type normal mouse fibroblasts to mutant transformed BHK cells, the grain count in the receiving cells was regularly higher than in the donor cells. How does that fit into the idea of a simple spread of nucleotides?

Pitts: With this radioautographic technique where acid-soluble material is lost, the grain count depends not only on the availability of labelled nucleotides but also on the extent of nucleic acid synthesis in the two cell types. It is therefore difficult to speculate about what this observation means.

Bard: I was intrigued by your suggestion that you might be pushing messenger RNA from one cell to the other rather than the enzyme itself. One can calculate that if the presumed enzyme had a molecular weight of about 50 000, the corresponding messenger RNA would have a molecular weight of about half a million and a length of about 1500 Å (150 nm). People who work with tight junctions reckon that you have reasonably free motility for molecules up to the order of 1000 Å and possible motility across the junction for molecular weights up to 20 000 or 30 000 (Furshpan and Potter 1968). A piece of string, as it were, 1500 Å long with a molecular weight of half a million seems to be a very different thing for which you probably need to invoke a different sort of communication system.

Dulbecco: These RNA's migrate in acrylamide gel electrophoresis, where the pores are very small; they can migrate lengthwise, displaying a small cross-section.

Pitts: Viral RNA can infect cells, but the mechanism of entry may be unrelated to the cell-to-cell transfer problem.

Bard: There is no evidence that I know of to suggest that viral RNA goes in via tight junctions.

Pitts: I drew attention to the possible relation between low-resistance junctions and cell-to-cell transfer, but transfer may not occur through such junctions. John Subak-Sharpe suggested that transfer occurs by exchange of cell membrane and cytoplasm between adjacent cells by a process akin to phagocytosis (Subak-Sharpe 1969). However, I don't think this is very likely, because transfer often occurs between cells which are touching only at a point or are joined only by a cytoplasmic bridge (Subak-Sharpe, Bürk and Pitts 1969). If it is by a process of mutual ingestion or phagocytosis I would expect it to happen predominantly between cells more extensively in contact.

Pontecorvo: You don't know the conformation the cells were in when they started to cooperate, however.

Pitts: In dilute culture, BHK cells don't tend to line up side by side, but tend to form an open network.

Warren: What do you think about Bendich's observation of cytoplasmic bridges between cells (Bendich, Vizoso and Harris 1967)?

Pitts: In a tissue culture all sorts of phenomena can be observed, but it is important to think in terms of how general they are. As yet, it isn't clear whether these bridges represent a normal feature of growing cells, or an abnormality found only in defective cells.

Shodell: If you want to show that tight junctions or low-resistance junctions are connected with this intercellular transport, one way of approaching it might be by the calcium sensitivity of low-resistance junctions. Possibly this could be done by seeing if there is a shutting down of this kind of transport between cells in contact, by reducing the calcium concentration transiently when you add the label and then raising it again in the absence of label.

Pitts: Yes; we haven't tried that yet.

Rubin: One alternative to the exchange of cytoplasm is the exchange of cell surface materials. It appears that if you label phospholipids in one cell and mix these cells with an unlabelled group of cells, the cells in contact with one another very rapidly exchange phospholipid (Peterson and Rubin 1970). This is apparently not true of proteins, but that doesn't mean very much because if there were some exchangeable protein in small amount in the cell surface, exchange might still happen and not be detected.

Pitts: We tried to develop a technique to identify mouse and hamster cells in mixed cultures. We made antisera against the two cell types and labelled them with fluorescein. We could easily distinguish between mouse cells and hamster cells in separate cultures, but in mixed culture all the cells appeared to fluoresce with either antiserum.

Rubin: That would indicate exchange of protein or carbohydrate.

Stoker: Could I bring this topic of cooperation more into the main area of the symposium and ask for comments on the way in which communication of this sort might be relevant to density dependent inhibition? In density dependent inhibition, we are dealing with only one cell type, but the physiology of the cells is heterogeneous; they are asynchronous as we study them and so I suppose one could consider that inhibition occurs through transfer of an inhibitor from cells that are in one stage of the mitotic cycle to those in another stage, so that the cells are in a communicating pool. If this were so, density dependent inhibition might not be effective in synchronous populations. Has anybody looked at contact or density dependent inhibition in synchronous cultures?

Bürk: I have studied 3T3 cells at different densities above or below confluence, putting them in 0·5 per cent serum, leaving them for four days to become quiescent and then starting growth by changing to fresh medium containing 10 per cent serum. Roughly speaking, at low density the time until incorporation of thymidine is shorter than at high density and there is less synchrony. There isn't a clear peak of incorporation and dividing cells appear earlier and are spread over a long time at low density. One could explain this by saying that at high density the cells can't behave independently and so they appear more synchronized, and on a simple model like Dr Pitt's model one would say that as TMP accumulates in one cell the nucleotide pool is depleted by being lost into neighbouring cells which aren't yet making it, so they have to wait for the last cell. The nucleotide pools are being increased at the beginning of DNA synthesis and they can't increase if they are leaking out into neighbouring cells which are not yet in S phase.

Pitts: This would affect the labelling but presumably not the synchrony of the cells, unless you are implying that the thymidine nucleotide pools regulate the induction of DNA synthesis?

Bürk: I think they are restricting it. If there are insufficient phosphorylated nucleotides, DNA synthesis doesn't go as fast or doesn't start at all.

Stoker: But you could imagine other small molecules affecting DNA synthesis?

Bürk: Yes. This is just one possibility. Cyclic AMP would be a more serious possibility.

Rubin: There is an argument against coupling by low-resistance junctions being the significant element in density dependent inhibition. Chick fibroblasts, which are very asymmetrical in shape, are almost all in contact with one another when only about a quarter of the surface of the dish is covered. At that point they are electrically coupled to one another; they establish coupling very quickly after contacting each other (P. O'Lague and H. Rubin, unpublished). Despite this functional contact the sub-confluent cells grow very rapidly, showing no sign of density dependent inhibition. It does not appear therefore that the mere existence of contact and the establishment of coupling affect the growth rate of cells. It is not until cells become confluent, and thereafter have increasing surface areas in contact with other cells rather than the substratum, that growth is inhibited.

Pitts: We get a similar result with radioautographic methods, looking at cell-to-cell interactions with BHK cells. They like to reach out and touch, and long before they reach confluence most of the cells in the culture will be interacting; they are growing quite rapidly at that point.

Rubin: At least as far as ions are concerned, which is what you measure when you measure coupling, their exchange reaches its maximal level very soon after cells touch (O'Lague and Rubin, unpublished); there is about as high a communication ratio then as when the cells are confluent. Even cells in mitosis that are communicating with other cells by very narrow strands are very nicely coupled together (O'Lague *et al.* 1970). Rous sarcoma cells are also coupled to each other, although the extent of surface contact may be small (O'Lague 1969).

Pitts: I am glad to hear that, because our growth experiments probably require cell-to-cell interactions throughout the cell cycle.

Weiss: The point is that coupling itself may not be the controlling mechanism but it would be interesting to know whether it is a prerequisite to growth control—whether there is no topoinhibition in cells which lack low-resistance junctions. The coupling means that some communication is possible, but maybe there are cells, like Rubin's Rous sarcoma cells, which are perfectly well coupled yet something inside the cell has lost the ability to respond to whatever the inhibitory influences are.

Stoker: But L cells grow to high density, don't they?

Weiss: Yes, and they lack metabolic cooperation; maybe they lack low-resistance junctions.

Clarke: Riley (1969, 1970), who has proposed a theory of growth control based on density dependent inhibition, finds that guinea-pig melanocytes will not divide if in groups of more than four cells; these cells might provide a useful tool for such an inquiry.

Weiss: There is evidence for the opposite effect in polyoma-transformed cells; if several such cells are in contact with each other this will overcome the topoinhibition of surrounding normal cells (Weiss and Njeuma 1971).

Bergel: Should one not distinguish, in the case of something passing from one cell to another, between charged and uncharged molecules? Charged molecules, whether small or large molecules (Bergel 1964), require special mechanisms to penetrate into the interior of a cell. Whether such processes are called pinocytosis, phagocytosis or endocytosis (see Jacques 1969), the most difficult event is the uptake of ionized compounds, such as nucleotides, including those studied by Dr Pitts. There is evidence that in contrast to the passage of purines into cells or their sugar derivatives (nucleosides), the phosphorylated compounds (acidic) have to overcome great resistance. This field of study can be divided into two: molecules necessary for cell growth or for inhibition of cellular activities either can easily be transported into the cell, or they cannot, under normal conditions. In the latter case, there appears to exist a special kind of contact mechanism between cells that forces the receptor cell to allow the uptake of charged particles or molecules, such as the nucleotides under discussion.

Pitts: It is unfortunate that our mutants all fall into a similar class. It would be nice to have a variety of mutants which we could use to examine the transfer of other types of molecules.

Dulbecco: There is a fundamental problem in the meaning of the electrical measurements. It could be that control does involve transfer of (large?) molecules between cells, but we cannot measure this transfer directly and we are forced to study the transfer of small ions. The measurements of conductivity between cells may be a poor way to study the ability of intercellular contacts to let through larger molecules. In fact, electrical conductivity may saturate very fast as pore size increases, reaching a maximum when pore size is still very small. Ultimately it is a question of the ratio of the resistance through the junction to the resistance of the total cell surface, and it may be that if resistance at the junction drops, say, from 1000 ohms to 1 ohm, that's enough for maximal ion movement and it makes no difference whether the resistance is now 1 ohm or 0·1 ohm. If this lower range is important for the transfer of larger molecules, then the study of electrical contacts is not a suitable method to detect regulatory differences in different cell types. It would be useful to know the minimal pore size at which resistance between cells reaches a minimum. In the past, pore size has been estimated by pushing substances through the pores by electrophoresis and seeing what goes through, but this has not been studied extensively. Measuring electrical contacts may be useless for studying regulation.

Elsdale: In all these experiments where one takes a confluent layer of cells of one type and puts a few cells of another type on top and sees if they form colonies, sometimes they do and sometimes they don't. Is there any correlation here between colony formation and electrical coupling between the two types of cells?

Weiss: Borek, Higashino and Loewenstein (1969) studied low-resistance junctions in heterologous contacts, between hepatoma cells and normal liver cells. There was no coupling and the growth of hepatoma cells was not inhibited by normal liver cells. Rous cells are not inhibited by normal cells either (Weiss and Njeuma 1971).

Rubin: The Rous cell is not inhibited by any cell but does exhibit electrical coupling.

Weiss: Are Rous cells coupled with cells of other types?

Rubin: I don't know.

Stoker: The SV40-transformed 3T3 cells are not inhibited by any other cells and they are coupled; do they couple with normal cells?

Dulbecco: This is not known.

Bard: The opposite information is perhaps more important: is there any cell which exhibits density dependent growth which is *not* coupled to its cellular environment? If you show that a cell is coupled but doesn't display density dependent growth it could mean merely that it is not capable of using the information it receives from outside.

Rubin: Some perfectly normal cells are coupled when grown in foetal calf serum but are uncoupled in calf serum (Borek, Higashino and Loewenstein 1969). The opposite is true for 3T3 cells and their transformed counterparts. There is no indication given that the state of coupling is correlated with the growth properties of the cells in this work.

Macpherson: Hasn't Loewenstein shown that coupling of cells is associated with the ability of dyes of different molecular weights to pass between the cells?

Rubin: He has done so with dyes varying in molecular weight from 300 to 1000 (Kanno and Loewenstein 1966). He also said that fluorescein-labelled serum albumin was transferred from cell to cell, but there is the danger that the fluorescein rather than the albumin was transferred. Polylysine of molecular weight 127 000 was not transferred. These experiments were done in *Drosophila* salivary glands, not in vertebrate cells.

Dulbecco: It has been suggested that this phenomenon of cooperation only occurs in cells infected with mycoplasma. Have you ruled out mycoplasma infection?

Pitts: We cannot tell whether our cells are free of mycoplasma; we do the tests and they are negative, but the cells may still be contaminated by undetectable levels of mycoplasma or by other infectious agents. However, several observations suggest that infection is not involved, for example: (1) BHK wild-type cells interact with mutant BHK cells but not with mutant L cells; (2) mutant BHK cells are phenotypically IPP⁻, APP⁻ and TK⁻, so selection in each case has given rise, not only to a cell mutant but also to the corresponding mutant of the infectious agent (if there is one); (3) the mutant phenotypes are very stable; (4) the ability of wild-type BHK cells to interact with BHK mutants is also a stable property; (5) the altered phenotype of BHK mutant cells in contact with BHK wild-type cells, however, is lost within minutes after separation; and (6) primary cells, which are less likely to be contaminated with infectious agents, also alter the phenotype of BHK mutants in contact with them. It seems very unlikely, therefore, that we are observing an infectious process.

Dulbecco: Ann Roman in my laboratory has done experiments similar to the one you described, by the dilution method, with normal 3T3 cells and 3T3 cells lacking thymidine kinase. In this system, we did not detect mycoplasma and we obtained exactly the same results as you did (unpublished results).

Stoker: Unpublished dilution experiments that Dr Pitts and I did also gave the same answer.

REFERENCES

BENDICH, A., VIZOSO, A. D. and HARRIS, R. G. (1967) *Proc. natn. Acad. Sci. U.S.A.* **57**, 1029.

BERGEL, F. (1964) *Spectrum International* **8**, 17, 33.

BOREK, C., HIGASHINO, S. and LOEWENSTEIN, W. R. (1969) *J. Membrane Biol.* **1**, 274–293.

FURSHPAN, E. J. and POTTER, D. D. (1968) *Curr. Top. Dev. Biol.* **3**, 95–127.

JACQUES, P. J. (1969) *Ciba Fdn Symp. Homeostatic Regulators*, p. 180. London: Churchill.

KANNO, Y. and LOEWENSTEIN, W. R. (1966) *Nature, Lond.* **212**, 629–630.

O'LAGUE, P. (1969) Ph.D. thesis, University of California, Berkeley.

O'LAGUE, P., DALEN, H., RUBIN, H. and TOBIAS, C. (1970) *Science* **170**, 464–466.

PETERSON, J. A. and RUBIN, H. (1970) *Expl Cell Res.* **60**, 383–390.

RILEY, P. A. (1969) *Nature, Lond.* **223**, 1382.

RILEY, P. A. (1970) *Nature, Lond.* **226**, 547–548.

SUBAK-SHARPE, J. H. (1969) *Ciba Fdn Symp. Homeostatic Regulators*, pp. 276–288. London: Churchill.

SUBAK-SHARPE, J. H., BÜRK, R. R. and PITTS, J. D. (1969) *J. Cell Sci.* **4**, 353–367.

WEISS, R. A. and NJEUMA, D. L. (1971) This volume, pp. 169-184.

ATTEMPTS TO ISOLATE SV40 TRANSFORMATION FACTOR

R. R. Bürk and Caroline A. Williams

Imperial Cancer Research Fund, London

TRANSFORMATION is a term for a group of changes in the growth of cells in culture which can occur either spontaneously or after infection by certain tumour viruses, or sometimes after exposure to carcinogenic chemicals. BHK21/13 baby hamster kidney cells can be transformed by polyoma virus (Stoker and Macpherson 1961) and less readily by SV40 virus (C. Wiblin, personal communication). 3T3 Swiss mouse cells can readily be transformed by SV40 virus (Todaro, Green and Goldberg 1964). The more obvious results of transformation of these cells by these viruses are the disorientation of the cells within the culture, the growth of the cultures in low concentrations of serum and the increased density to which the cultures grow. It seems possible that transformed cells may make a cascade of one or more substances which act to produce the syndrome of transformation. We would like to isolate one of these substances.

An intrinsic difficulty is that these substances are produced in the cell and act in the cell, whereas any substance isolated will be outside the cell. A crude approach is to hope that if there is enough outside the cell some will get in. Alternatively if one of the substances does act from outside the cell, it should be released by the cell into the medium. Our approach, then, has been to prepare extracts of medium which has been exposed to SV40-transformed cells. A BHK21/13 cell line transformed by SV40 virus, called SV28, was chosen because it grows to high density in culture itself and because BHK21/13 also grows to high density, allowing comparisons at high density. In principle, comparisons between BHK21/13, 3T3 and their SV40 and polyoma-transformed derivatives should be an approach to recognizing hamster, mouse and virus gene products.

METHODS

Cells

The baby hamster kidney cell line, BHK21/13, and a polyoma virus-transformed derivative PyJ, were obtained from Dr M. G. P. Stoker.

SV28, an SV40 virus-transformed derivative of BHK21/13, was obtained from C. Wiblin. The Swiss mouse cell line, 3T3, is our derivative of clone 14 of Dr R. W. Holley, and its SV40 virus-transformed derivative, SV3T3, was obtained from Dr R. Dulbecco.

Culture

Cells were maintained by normal techniques in the Dulbecco modification of Eagle's medium (9 parts) supplemented with (1 part) calf serum (E_9C_1).

For the preparation of extracts, cells were grown in roller bottles with 200 ml of (Eagle's medium, 9 parts; tryptose phosphate broth, 1 part; calf serum, 1 part: $E_9T_1C_1$) until the cultures were dense ($2–4 \times 10^8$ cells). The medium was replaced by 200 ml (Eagle's medium, 9 parts; tryptose phosphate broth, 1 part: E_9T_1) and collected after 24 hours.

Preparation of extract

The medium was placed on ice for 30 minutes, solid KCl was added to 2 M and the medium was kept on ice for 30 minutes more. Two volumes of ethanol were added and after half an hour on ice to give a temperature of 6°C, the precipitate was centrifuged down at 4°C and discarded. The same volume of ethanol as previously was added to the supernatant (giving about 80 per cent ethanol) and after two hours on ice the precipitate containing much KCl was centrifuged down, resuspended in 10 ml phosphate buffered saline (PBS) per 200 ml of starting medium and dialysed against 12–50 volumes of PBS. After an hour the dialysis bags were changed to fresh PBS and the next day to water. After one day's dialysis against water the precipitate was centrifuged down and suspended in (0·1 M-acetic acid + 0·05 M-NaCl, pH 2·75).

Gel chromatography

The column had a bed-volume of 112 ml, an internal diameter of 17 mm and was packed with Sephadex, G75 medium.

Assays

G_2-mitosis. $1·0 \times 10^6$ BHK21/13 cells were seeded in 5 ml E_9C_1 medium in 5 cm plastic Petri dishes. At 3 hours the medium was sucked off and replaced by $(E_9T_1)_{400}C_1$—that is, 0·25 per cent serum. At 21 hours the extracts were added to pairs of dishes and at 26 hours the cells were removed with a trypsin/EDTA/PBS mixture and counted in a Model B Coulter Counter.

Overgrowth. 5×10^5 3T3 cells were seeded in 5 ml E_9C_1 medium in 5 cm plastic Petri dishes. After 3 hours the extracts were added and on the third day the number of cells was determined as above.

Low serum. $1 \cdot 0 \times 10^6$ BHK21/13 cells were seeded in 5 ml $E_{50}C_1$ medium in 5 cm plastic Petri dishes. After 3 hours the extracts were added and on the third day the number of cells was determined as above.

Thymidine incorporation. $1 \cdot 25 \times 10^6$ 3T3 cells were seeded in $2 \cdot 5$ ml $E_{200}C_1$ medium in 35 mm plastic Petri dishes containing two 12 mm diameter glass coverslips. After four days the extracts were added and 21 hours after the extracts, $0 \cdot 02$ ml of 250 μM-thymidine (methyl-^{3}H) (100 μCi/ml) was added. At 27 hours the medium was sucked off, the plates washed with tris-saline and the cells fixed in formol saline. The coverslips were then washed in 5 per cent trichloroacetic acid at $0°$ C, twice in water at $0°$ C, then in ethanol, air-dried and placed in $0 \cdot 5$ per cent PPO (2,5-diphenyloxazole) in toluene and the radioactivity determined in a Nuclear Chicago Mk II Liquid Scintillation Counter with a 50–55 per cent efficiency (Bürk 1970).

Uridine incorporation. The procedure was as for thymidine except that $0 \cdot 02$ ml of 250 μM-uridine (5-^{3}H) (100 μCi/ml) was added at the same time as the extracts on the fifth day and the cells were fixed after 3 hours.

RESULTS

The origin of the extraction procedure is only of historical interest and cannot be described here. Different assays have been used, each having different merits and disadvantages, and these will be discussed in some detail.

The *uridine incorporation* assay takes about four hours and being so rapid was used at first. Table I shows the results when the procedure

TABLE I

EFFECT OF EXTRACTS OF MEDIUM CONDITIONED BY 3T3 AND SV3T3
CELLS ON URIDINE INCORPORATION BY QUIESCENT 3T3 CELLS

Extract	Replicate medium	Percentage uridine incorporation*
SV3T3	A	169
	B	151
3T3	A	97
	B	84

* 100 per cent= 1280 counts/min in control.

was taken as far as the second ethanol precipitate. The activity in the extract of SV3T3 medium was independent of the uridine concentration used in the assay, suggesting that the effect was not on uptake of uridine but on RNA synthesis. However, the paper of Cunningham and Pardee (1969) demonstrated the weakness of this assay.

The G_2-*mitosis* assay takes only about six hours and was preferred to a mitotic index assay because an increase in cell numbers is an unambiguous measure of growth whereas a rise in the mitotic index can result from slower mitosis. The G_2-mitosis assay is intrinsically inaccurate. The 3-hour pulse of high serum (E_9C_1) stimulates a growth cycle in 75 per cent of the cells at most and if the timing is not quite right some cells will have already divided in the control, giving a high blank and reducing the possible increase in cell number still further. Fig. 1 illustrates a dose-response curve for this assay. The advantage of the assay is that it is not affected by whole serum. Table II shows the results of the G_2-mitosis assay

TABLE II

EFFECT OF SERUM, CONTROL BUFFER AND EXTRACTS OF MEDIUM IN
G_2-MITOSIS ASSAY: YIELD AT 26 HOURS IN MILLIONS OF CELLS

Treatment	Volume added ml	Extracts of replicate cultures	
Count at 21 hours	—	0·90	0·92
Blank	—	1·00	1·06
Continuous serum	0·2	1·57	1·51
Serum at 21 hours	0·2	1·01	1·06
Control buffer	0·1	0·96	1·06
	0·02	1·01	1·01
SV28 medium extract	0·1	1·19	1·18
	0·02	1·11	1·14
BHK21/13 medium extract	0·1	1·07	1·10
	0·02	1·00	1·07

when the extraction procedure was taken as far as the PBS precipitate which was then extracted in 0·1 M-acetate, pH 4·0. Continuous serum gives a measure of how many cells could divide and the count at 21 hours indicates how many cells were treated with the extract. Serum added at 21 hours had no effect. It can be seen that 0·1 ml of the BHK21/13 extracts had a small effect and that 0·02 ml of the SV28 extracts had a greater effect than 0·1 ml of BHK21/13 extract, indicating more than five times as much activity in the SV28 extracts.

The *overgrowth of* 3T3 assay (continuation of growth in dense cultures) has proved to be the most reproducible and accurate assay. It is not very convenient. The growth characteristics of 3T3 make it difficult to have large numbers of these cells readily available and the assay takes three days.

The assay is done in 10 per cent serum and so is expected to be unaffected by small amounts of serum. Fig. 2 illustrates a dose–response curve for

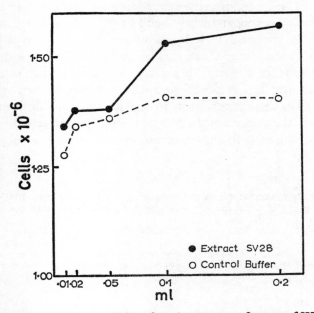

FIG. 1. G₂-mitosis assay. Effect of varying amounts of extract of SV28 medium.

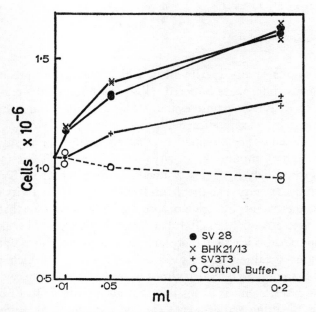

FIG. 2. Overgrowth of 3T3 assay. Effect of varying amounts of extract of SV28, BHK21/13 and SV3T3 medium.

this assay. The best extracts have produced a 140 per cent increase in cell numbers. Occasionally when the 3T3 cells have been "spontaneously" transformed they seemed no longer sensitive. Some extracts of BHK21/13 medium have less activity than parallel extracts of SV28 medium but on other occasions the activities of extracts of these two media are similar. There seems to be sufficient agreement between extracts of replicate cultures within these experiments to conclude that the differences reflect differences in the cultures of the cells and not in the extraction.

The desalting step in the extraction seems rather critical. Table III

TABLE III

EFFECT OF DESALTING ON ACTIVITY OF EXTRACTS OF SV28 AND BHK21/13
MEDIUM IN 3T3 OVERGROWTH ASSAY: PERCENTAGE INCREASE IN CELL NUMBERS

Extract	Replicate medium	Dialysis precipitate		Dialysis supernatant	
		$0 \cdot 01$ ml	$0 \cdot 1$ ml	$0 \cdot 01$ ml	$0 \cdot 1$ ml
SV28, PBS	1	8	44	4	55
	2	10	55	10	54
SV28, H_2O	3	29	76	1	17
	4	25	72	−2	4
BHK21/13, PBS	1	10	40	3	37
	2	16	57	10	42
BHK21/13, H_2O	3	6	33	−2	8
	4	17	59	−1	6

PBS, phosphate buffered saline.

shows the results of an experiment in which the alcohol precipitate was dialysed against phosphate buffered saline and some of the extracts were then further dialysed against water. The activity in the supernatant was recovered by precipitation with four volumes of ethanol. A comparison of the activities obtained by dialysis against saline suggests that there is the same amount of activity in SV28 and BHK21/13 medium but after further dialysis, against water, the precipitate from SV28 is seen to contain more activity than that from BHK21/13. The activity recovered from the supernatant by ethanol precipitation indicates that there is fairly complete precipitation after dialysis against water. This experiment might support the idea that there are two substances in SV28 extracts with different desalting characteristics, only one of which is in the BHK21/13 extracts. So far we have not been able to separate two activities from SV28 extracts and it is possible that all the differences observed in the various desalting experiments can be explained by differences in concentration of the one active substance in the extracts.

Extracts have also been prepared from 3T3 and SV3T3 medium. The 3T3 extracts have (Table IV) no activity whereas SV3T3 extracts

TABLE IV

ACTIVITY OF 3T3 AND SV3T3 MEDIUM IN 3T3 OVERGROWTH ASSAY: PERCENTAGE INCREASE IN CELL NUMBER

Extract	Replicate medium	Dialysis precipitate		Dialysis supernatant	
		0·05 ml	0·2 ml	0·05 ml	0·2 ml
3T3	1	1	1	5	16
	2	6	1	9	25
SV3T3	1	28	40	34	52
	2	28	45	33	55

have activity. These extracts were prepared by dialysis against phosphate buffered saline. It should be remembered that the 3T3 cultures are much less dense than the SV3T3 and are non-growing. Ethanol precipitation of the supernatant, as above, indicated some activity in the extracts of 3T3 medium. Activity in SV3T3 extracts has always been less than in SV28 extracts.

We tried to determine whether the activity in the extracts of SV28 culture media comes from the calf serum in which the cells were grown. The hypothesis was that SV28 cells took up serum in some way and when the cultures were changed to medium lacking serum the cells released the serum component. Extracts were therefore prepared from fresh $E_9T_1C_1$ medium, fresh E_9T_1 medium and $E_9T_1C_1$ medium incubated in a culture bottle without cells, and from E_9T_1 medium from SV28 cells as a positive control. As can be seen from Table V there was less than 5 per cent of SV28

TABLE V

ACTIVITY OF DIFFERENT MEDIA IN 3T3 OVERGROWTH ASSAY: PERCENTAGE INCREASE IN CELL NUMBER

Medium	Replicate medium	Volume added, ml		
		0·005	0·02	0·1
SV28 E_9T_1	1	31	69	92
	2	25	64	90
$E_9T_1C_1$ fresh	1	—	—	17
	2	—	—	12
$E_9T_1C_1$ incubated	1	—	—	17
	2	—	—	20
E_9T_1 fresh	1	—	—	0
	2	—	—	−7

activity in the extracts of media containing serum, so it seems that cells are necessary for the production of the activity. It is not known whether the active substance is synthesized by the cell or whether the cells "activate" a substance in serum.

The *low-serum BHK21/13* assay is based on the observation (Fig. 3) that BHK 21/13 cells grow more slowly in 1 per cent serum than in 10 per cent, whereas SV28 cells grow at similar rates. It was found, however (Fig. 4), that extracts of SV28 medium were inhibitory in this assay. The cells in the inhibited BHK21/13 cultures were disoriented (compare

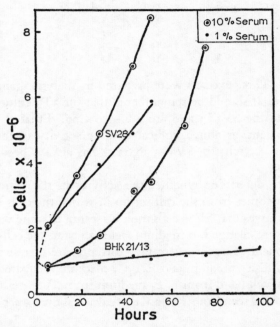

FIG. 3. Growth of BHK21/13 and SV28 cells in Eagle's medium supplemented with 1 per cent and with 10 per cent calf serum.

the upper and lower micrographs of Fig. 5). This inhibition was also obtained in 10 per cent serum. These results suggest that the extracts contain an overgrowth stimulator for 3T3 and a growth inhibitor for BHK21/13, or that the one substance acts differently on 3T3 and BHK21/13 cells—that is to say, the cells are different. The purification of the activities should settle this point. When an extract of SV28 medium was fractionated at pH 2·75 in 0·1 M-acetic acid+0·05 M-sodium chloride on G75 Sephadex (Fig. 6) it was found that the 3T3 overgrowth activity was eluted later than the bulk of the protein (material with absorbancy at 280 nm) in a reasonably discrete peak. The major part of the inhibitory activity in the low-serum BHK21/13 assay was eluted in the same position as the overgrowth activity, followed by a later peak. The same fractions which were inhibitory to BHK21/13 were also inhibitory to the growth of whole mouse embryo tertiary cells in 2 per cent serum. In the desalting experiments also, the BHK21/13 inhibitory activity corresponded with the 3T3 overgrowth activity. So far the 3T3 overgrowth activity and the inhibitory activity for low-serum BHK21/13 have not been separated.

When SV28 extracts were fractionated on G75 Sephadex at pH 2·75

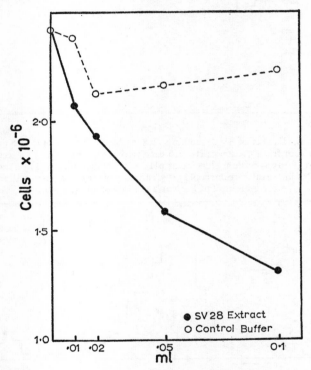

Fig. 4. Low-serum growth of BHK21/13 assay. The effect of varying amounts of extract of SV28 medium.

with bovine serum albumin and ribonuclease A they behaved as though the 3T3 overgrowth activity were associated with a substance of molecular weight 27 000 (Fig. 7). The 3T3 overgrowth activity in BHK21/13 extracts behaved in a way indistinguishable on G75 Sephadex from the activity from SV28 medium.

Further purification of the substance with activity in the 3T3 overgrowth assay is clearly necessary. However, it seems desirable to examine the properties of the substance from SV28 medium obtained by fractionation at pH 2·75 on G75 Sephadex.

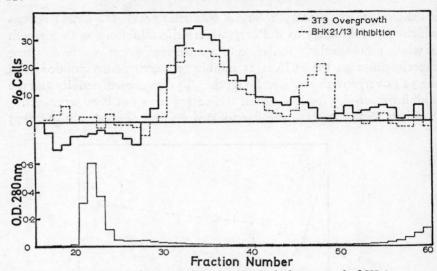

FIG. 6. Extract of SV28 medium on G75 Sephadex. 200 ml of SV28 medium from 3×10^8 cells were extracted into 3 ml of 0·1 M-acetic acid + 0·05 M-sodium chloride at pH 2·75, applied to the column, and eluted in 1·7 ml fractions at pH 2·75, and assayed for 3T3 overgrowth and for inhibition of BHK21/13 in low-serum growth.

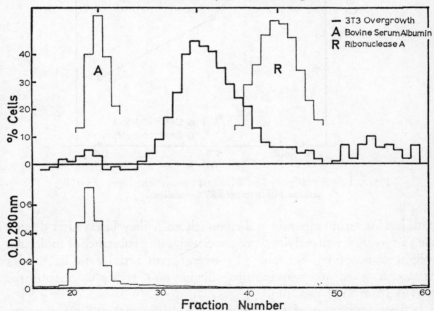

FIG. 7. Extract of SV28 medium on G75 Sephadex compared with bovine serum albumin and pancreatic ribonuclease A. 400 ml of SV28 medium from 7×10^8 cells were extracted into 3 ml of 0·1M-acetic acid + 0·05M-sodium chloride (pH 2·75) and applied to the column (for details see Fig. 6).

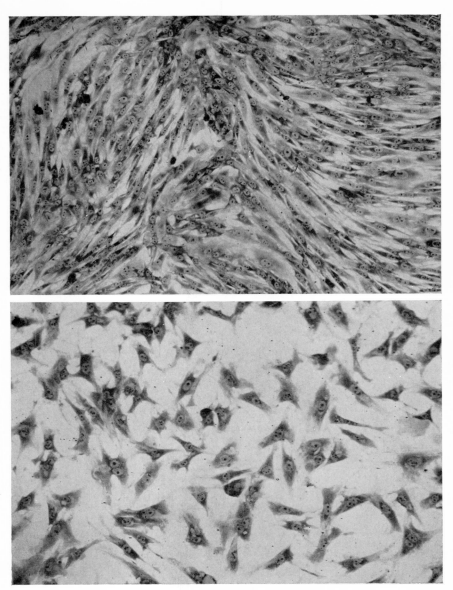

Fig. 5. Effect of extract of SV28 medium on the orientation of BHK21/13 cells. The cells were seeded at 2×10^5 per 5 cm plate in 2 per cent serum medium and after 3 hours $0 \cdot 1$ ml of control buffer (*upper*) or $0 \cdot 1$ ml of the extract of SV28 medium (*lower*) was added. After 4 days cells were fixed with acetic acid/ethanol and stained with Giemsa. $\times 150$.

[*To face page 116*

When the Sephadex G75 fractions were tested in the [³H]uridine incorporation assay those fractions with most 3T3 overgrowth activity had a small (up to 20 per cent) stimulating effect on the incorporation of [³H]uridine.

When the fractions were tested in the [³H]thymidine incorporation assay there was a broad peak about 2·5 times the background (350 counts/

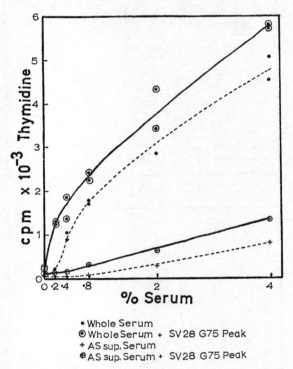

FIG. 8. Enhancing effect of extract of SV28 medium on stimulation of thymidine incorporation by whole calf serum and by the ammonium sulphate supernatant fraction of serum.

min) corresponding to the 3T3 overgrowth activity. All the fractions were also assayed with 0·01 ml (0·4 per cent) calf serum added to each plate. The peak was in the same place and was still 2·5 times the background (1800 counts/min) but somewhat smoother. The fact that in the first assay the peak stimulation was about 475 counts/min and in the assay with added serum, about 2700 counts/min suggests that the fractions are not stimulating the incorporation of thymidine themselves but are enhancing or potentiating the effect of serum. This was tested by measuring the activity of pooled G75 peak fractions in the [³H]thymidine incorporation assay on 5-day quiescent 3T3 cells with varying amounts of whole calf serum and varying amounts of calf serum from which much of the activity had been removed by treatment with 40 per cent saturated ammonium sulphate (Jainchill and Todaro 1970). The results in Fig. 8 show that the peak fractions from SV28 medium do not themselves stimulate thymidine incorporation but

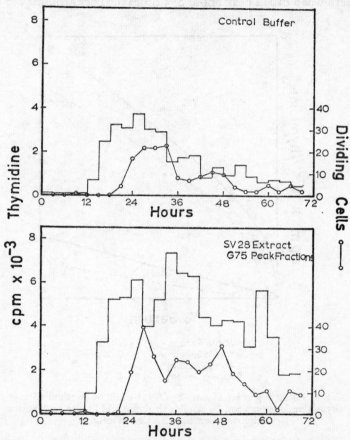

FIG. 9. Overgrowth effect of the extract of SV28 medium on quiescent 3T3 cells stimulated by 2 per cent whole calf serum.

that in their presence the incorporation was equivalent to about the incorporation in their absence with twice as much serum. They seem to double the effect of the serum on 3T3 cells. Finally, we examined the effect of pooled G75 fractions from SV28 medium on the stimulation of growth of quiescent 3T3 cells at high density by a limiting amount of whole calf serum. $2 \cdot 5 \times 10^5$ 3T3 cells were set up as in the thymidine incorporation assay in $0 \cdot 5$ per cent serum. After four days, $0 \cdot 05$ ml serum was added to all plates (i.e. 2 per cent serum). Half the plates received 0·05 ml of control buffer and half received 0·05 ml of pooled G75 peak fractions. Thymidine incorporation and the relative number of dividing cells in 3-hour periods was determined (Bürk 1970). In the control (Fig. 9) a rather broad peak of thymidine incorporation and cell division was seen with not much activity after the first peak. In the treated plates thymidine incorporation was almost doubled in the first rather more synchronous peak. There was a second higher peak of thymidine incorporation with no or very little G1 phase and probably even a third round of thymidine incorporation. The first round of cell division was also quite well synchronized at about 27 hours and was followed by much more cell division with a peak at 48 hours. This experiment agrees with the potentiating effect observed in the previous experiment and may indicate a speeding of the cell through S and G_2–mitosis.

DISCUSSION

The main findings reported in this paper are that extracts prepared from the media of BHK21/13 and SV28 cells can stimulate the overgrowth of 3T3 cells. Extracts of SV3T3 medium had less activity than those of BHK21/13 and SV28 media. Extracts of 3T3 medium had very little activity. The activity was not obtainable from serum in the absence of cells. Extracts active in the 3T3 overgrowth assay were inhibitory to the growth of BHK21/13 in 2 per cent serum medium, producing disorientation of the cells. The overgrowth and inhibitory activities of SV28 extracts were not separated on G75 and behaved as if they corresponded to substances of molecular weight of 27 000.

Our objective has been to isolate an SV40 transformation factor. It is clear from the activity and the optical density at 280 nm of the fractions from the G75 Sephadex column that the active substance(s) has not been isolated. Some measure of progress is indicated by a calculation which assumes that the substance is protein and has a molecular weight of 27 000, assumes 20 μg/ml protein in the extract, and takes 5 μl as active on 10^6 cells. This calculation indicates that fewer than 10^6 molecules per cell can produce an effect in the assay. Hopefully, further purification would lower this estimate. There is some activity in BHK21/13 extracts in both the 3T3 overgrowth assay and the BHK21/13 low-serum growth assay, so the simplest model is that the active substance is a cell product, not a virus product. Notice, however, that the active substance has not been shown to be the same in the extracts from different media. The ability of a substance to allow 3T3 cells to grow to a high density seems to be a character required of a transformation factor. The serum-potentiating effect for thymidine incorporation into 3T3 is also consistent with the lower serum requirement of polyoma and SV40-transformed cells (Stanners, Till and Siminovitch 1963; Bürk 1966; Holley and Kiernan 1968). The inhibitory activity in the low-serum BHK21/13 growth assay and the whole mouse embryo tertiary cell growth assay so far behaves as though it is the same substance as the one which stimulates 3T3 cells. This activity does not seem to be expected for a transformation factor. The disorientation of the BHK21/13 cells has not been quantified and so it has not been possible to be sure that the fractions with a maximum disorientating effect corresponded to the fractions with the maximum effect in the other assays, but it looked that way. We hope that further purification will resolve these problems.

SUMMARY

There may be substances in cells transformed by SV40 virus, coded by the cell or the virus, which act to produce all or some of the changes in cell properties known as transformation. While such substances may only act from within a cell, if they acted from without they might be found in the culture medium. Extracts have been prepared from culture medium of 3T3 and SV3T3 cells and of BHK21/13 and (BHK21/13)SV28 cells and assayed for stimulation of uridine or thymidine incorporation into quiescent 3T3 cells, for stimulation of overgrowth of 3T3 cells, for stimulation of growth of BHK21/13 in low serum and stimulation of the cell cycle of BHK21/13 in low serum, and for stimulation of the cell cycle of BHK21/13 in G_2-mitosis. A factor has been found in SV28 medium which behaves on Sephadex filtration at pH $2 \cdot 7$ as though it is a substance of mol. wt. 27 000 which stimulates 3T3 overgrowth at concentrations of less than 1 μg/ml protein. There is similar activity in extracts of BHK21/13 medium and SV3T3 medium and slight activity in extracts of 3T3 medium. Extracts of SV28 medium are stimulatory in the BHK21/13 G_2-mitosis assay but are inhibitory in the BHK21/13 low-serum growth assay, producing disorientation of the cells.

REFERENCES

Bürk, R. R. (1966) Nature, Lond. 212, 1261–1262.
Bürk, R. R. (1970) Expl Cell Res. 63, 309–316.
Cunningham, D. D. and Pardee, A. B. (1969) Proc. natn. Acad. Sci. U.S.A. 64, 1049–1056.
Holley, R. W. and Kiernan, J. A. (1968) Proc. natn. Acad. Sci. U.S.A. 60, 300–304.
Jainchill, J. L. and Todaro, G. J. (1970) Expl Cell Res. 59, 137–146.
Stanners, C. P., Till, J. E. and Siminovitch, L. (1963) Virology 21, 448–463.
Stoker, M. and Macpherson, I. (1961) Virology 14, 359–370.
Todaro, G. J., Green, H. J. and Goldberg, B. (1964) Proc. natn. Acad. Sci. U.S.A. 51, 66–73.

DISCUSSION

Shodell: You find a small amount of overgrowth activity in medium from cultures of normal cells; could that indicate that it is not a virus-coded or virus-induced activity, but simply that transformed cells leak more of the active substance into the medium? Such differences in the initial concentration of the material could also result in different precipitation profiles.

Bürk: Yes, it probably is a cell growth factor, since we get similar amounts of activity from BHK and SV40-transformed BHK (Fig. 2, p. 111). We even had the suggestion earlier in the meeting (p. 31) that it might be a decay product from the cells; you could say that under our growth regime the normal cells survive better than transformed cells and don't decay, and that is why sometimes when they are healthy the factor isn't released into the medium. We have tried making and testing extracts of cells, which should answer this point, but one starts with such a mush of all sorts of things that the extraction can't be said to be the same. We get out activity that is of the same order of magnitude as we get from the medium, but there are many biochemical problems here.

Shodell: Is the extract inhibitory to BHK cells under a range of serum conditions?

Bürk: We have done only 2 per cent and 10 per cent serum; the extract is more inhibitory at 2 per cent than at 10 per cent, but the problem is that BHK cells grow more at 10 per cent. For instance there are eight million cells in the control with 10 per cent serum, and a decrease of two million cells with extract; whereas there are three million cells in the control with 2 per cent serum and a decrease of one and a half million cells with extract. It is a different percentage of the control and I don't know what units to express it in.

Stoker: We measured thymidine incorporation in medium with no serum and with 0.5 per cent serum and the Bürk extract is inhibitory at 0.5 per cent and also inhibits the small amount of background incorporation that you see without serum.

Shodell: It might also be useful to normalize the serum concentrations to the particular cell type used in the assay. That is, at the same absolute serum concentration you are really dealing with a higher effective serum concentration for 3T3 cells than for BHK, since the former would react much more sensitively to such serum changes in an overgrowth assay.

Holley: It would depend on the amount of serum you had, but certainly 3T3 cells are more dependent on serum than BHK cells.

Shodell: The related question would be, does the extract give overgrowth of 3T3 even at low serum concentrations?

Bürk: I don't know about overgrowth. The only time we varied the serum concentration was in the thymidine incorporation experiments.

Cunningham: I understand that when you extracted the medium from 3T3 cells, they were confluent. Has it been possible to assay for activity in extracts of medium from non-confluent 3T3 cells?

Bürk: We haven't done this; it would be difficult to get enough cells.

Dulbecco: Can we make a comparison between the various factors that have been discussed? Dr Holley gets his from serum and Dr Stoker gets his from BHK cells; to what extent are these extracts similar?

Holley: There is similarity in chromatography between Dr Bürk's factor and ours; there is a dissimilarity in stability between our factor and Dr Stoker's—his is much less stable.

Stoker: The other difference between the activity we find in the medium and Dr Bürk's extract is that our material stimulates BHK cells and his inhibits them.

Holley: Have you tried other methods to separate the inhibitory activity from the stimulating activity, such as heat stability or pH stability?

Bürk: We haven't tried heating the extract. We have concentrated on trying to separate out the overgrowth activity, and despite varying a number of things we get the inhibitory activity with it.

Holley: Is the overgrowth activity stable if you neutralize the extract and then heat it?

Bürk: We started by dialysing it and then by trying to extract the precipitate at higher pH, above pH 7, and we found no overgrowth activity; we lost all activity when we dialysed against pH 8·4, but we haven't done the experiment of neutralizing it and seeing if it is stable.

Bard: I was wondering what you meant by "inhibition" of the BHK cells. In your micrographs (Fig. 5) of the healthy cells and those after addition of inhibitor, the cells in the latter case just looked rather sick. The solution seems to act in a manner more toxic than inhibitory. When you took the extract off, did the cells look healthy?

Bürk: I haven't tried taking the extract off.

Dulbecco: These various growth factors all need small amounts of serum for assay. What is the role of serum? Is it needed to protect some of the growth factors themselves which are labile, or does it have a more specific effect? Can one replace the serum background with other proteins or with something else which might have a stabilizing effect?

Holley: Certainly not any protein will work; on an electrophoretic pattern one finds certain fractions that lead to survival of 3T3 and SV3T3 cells and are additive with other things, and other fractions which are not. I hope that in the long run we shall have purified substances.

Taylor-Papadimitriou: Which fraction is effective in promoting survival?

Holley: "Survival" activity comes in the albumin region but it is not a sharp peak.

Burger: Dr Holley, has it been shown or investigated whether any of these components of serum that give additive growth effects are effective as stabilizers from decay in the absence of cells? At neutral pH, these purified fractions are not very stable. What is the survival time of a combination of all these factors at say 37°C in the absence of cells in various dishes like glass, plastic and so on?

Holley: Added to the medium and then put in the incubator (37°C), they seem quite stable, judging by our assay. This instability that I spoke of is probably a loss on the surface of the tube or precipitation, rather than an actual chemical instability.

Warren: To go back to the possible origins of materials that serve as factors of one sort or another, Dr Bürk mentioned "decay" products again as possible factors. I would like to put this on a more rigorous basis and introduce the question of the metabolic turnover of the cell and especially of its surface membrane. Dr Glick and I showed (Warren and Glick 1968) that when L cells were not dividing there was a very large turnover of surface membrane, measured by incorporation of isotopic precursors into surface membrane and also by the rate of loss of label from the membrane. We employed radioactive glucosamine, amino acids and glucose and measured radioactivity in the chloroform-methanol-soluble fraction, which more or less covers the whole membrane—carbohydrate, protein and lipid. The labelling patterns of each of these were roughly in parallel. We found that the cell makes as much surface material when it is not dividing as when it is dividing; when the cell divides, the new material goes into the formation of the new surface membrane required. As soon as the cell stops dividing it goes out of gear; the new material is synthesized at the same rate and goes into the membrane but an equivalent amount has to go out and that is where the turnover occurs.

This may be relevant to two topics that have been discussed here: one is the formation of factors that enter the serum which may have some relationship to the cell surface in terms of derivation, composition and affinity for cells in culture. The other has to do with intercellular junctions. If two cells with their surface membranes turning over rapidly approach each other, they might have a harder time forming a junction of any kind than if the two are not turning over. So, metabolic turnover has to be considered in both situations. It is not simply a random decay of the cell; it is an integral and certainly a regulated process. It would be extremely interesting to know when Dr Bürk's factors are formed; are they formed in log phase or plateau? I would predict that they would be formed in plateau when there is turnover and at least some of this rejected material must be entering the medium. When we fed cultures with [^{14}C]glucosamine, which is the precursor of N-acetylglucosamine, N-acetylgalactosamine and sialic acid, we found macromolecular material in the medium that contained labelled sialic acid. We like to think it is rejected surface membrane, and it may suggest the possible origins of the various factors discussed, which are perhaps not unlike lipoprotein materials, with more or less affinity for lipids, that are found normally in serum. Serum may be filled with materials that have their origins in the surface membranes of cells and may play an important part in the control of cell growth.

Bürk: My use of the word "decay" is more applicable to our cells, which when deprived of serum and left for a few days all disintegrate and are in a very bad state so that when we harvest them after one day, they are on this road down. I take your point, however, that there should be turnover as well; the decay may be accelerated turnover.

Stoker: The products of turnover of membranes will only be relevant, of course, in the systems we are talking about if they have some biological activity when they reattach to cells.

Montagnier: Dr Warren's work on the turnover of cell membrane material was done on cells growing in suspension. This is a special situation which is different from that of monolayers.

Warren: When we grew these L cells as monolayers we got essentially the same turnover results.

Montagnier: Have you tried other types of cells?

Warren: No. We measured the rate of incorporation of various isotopes into cells before and after viral transformation, with Dr V. Defendi. We were asking the question, what would be the consequences of high and sustained turnover; could this be the basis for the manifestation of malignancy, that the malignant cell keeps going, turning over at a high rate, whereas a normal cell stops both division and turnover? We have got nowhere with this; we found that some transformed cells had twice the turnover rate of their non-transformed counterparts, but others had the same rate (see Warren 1969).

Montagnier: It might depend on the medium and culture conditions.

Dulbecco: Dr Warren, the question of contamination with PPLO is serious in your experiment, because if it is present, the turnovers you measure are of PPLO.

Warren: We have always checked this. We grow the cells in aureomycin once a week, and our samples are checked in the Department of Bacteriology every month. We have also used several different preparations of

L cells. We have looked at the turnover of sialic acid among other things, and it behaved just like the other components studied. PPLO does not contain sialic acid.

Shodell: In the conditioning effects that Harry Rubin was working with some years ago, we found that chicken serum could be considered stimulatory for cultures of primary chick cells, that calf serum was inhibitory and that the conditioning factor did in some way interact with calf serum to alleviate its negative effect and thereby conditioning became stimulatory as well. This conditioning effect in chick cells moved during fractionation with cell surface material which was produced into the medium; it was identified as surface material by electron microscopy and a limited amount of chemical characterization. So there is some evidence from these experiments for surface material which is produced into the medium and which interacts with serum to stimulate the growth of chick cells.

REFERENCES

WARREN, L. (1969) *Curr. Top. Dev. Biol.* **4**, 197.
WARREN, L. and GLICK, M. C. (1968) *J. Cell Biol.* **37**, 729.

GROWTH REGULATION IN CULTURES OF CHICK EMBRYO FIBROBLASTS

H. RUBIN

Department of Molecular Biology and Virus Laboratory, University of California, Berkeley, California

TISSUE culture studies of cell growth regulation have the advantage over *in vivo* studies of having a manageable number of variables which can influence the growth rate and are accessible to manipulation and measurement. This paper is a survey of some of these variables derived from accumulated unpublished data of experiments with chick embryo fibroblasts done over a number of years. I restrict myself to data from a single system to provide a concrete and immediate base for our discussion of possible mechanisms of growth regulation and for comparison with other systems.

All the experiments to be described were done with primary or secondary cultures of cells from 10-day-old chick embryos. The cells were trypsinized and grown in 60 mm Petri dishes according to a standardized procedure (Rein and Rubin 1968). The growth medium, unless otherwise stated, consisted of medium 199 with tryptose phosphate broth (TPB) and chicken serum. The growing cell population consisted almost entirely of chick embryo fibroblasts (CEF).

THE SERUM REQUIREMENT

The growth rates of CEF seeded at different cell densities were roughly proportional to the serum concentration (Fig. 1). There was an almost stoichiometric relation between the concentration of cells present and the minimal serum concentration required to give the fastest growth. This relation continued as the cell population increased with time, so that the serum requirement increased, although there was no evidence for depletion of the growth-promoting activity of the serum at any time.

Many commercial sera, when used at high concentration, are toxic to CEF, particularly when the population density of the cells is low. Calf sera are in general more toxic than chicken sera (Fig. 2). Fresh chicken sera obtained by bleeding living chickens are not toxic to CEF, even when the ratio of serum to cells is high, and such sera were used in all experiments where small numbers of cells were grown. When possible the sera were obtained from a flock with the same genetic background as that of the embryos used to provide the cells for culture.

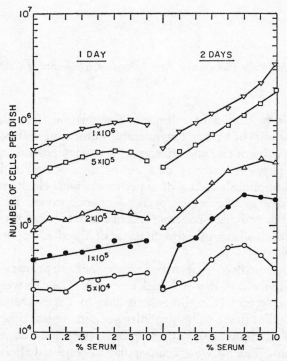

FIG. I. Graph of chicken serum concentration against cell concentration. CEF were seeded at various concentrations in 60-mm plastic Petri dishes in medium 199+2 per cent tryptose phosphate broth (TPB)+the chicken serum concentration shown in the abscissa. The number of cells seeded is shown under the appropriate curve in the left-hand panel. Cells from two dishes were trypsinized at 1 and 2 days and counted in the Coulter Counter.

Some characteristic features of the kinetics of growth of CEF seeded at various cell concentrations in various concentrations of a non-toxic chicken serum are illustrated in Fig. 3. Cells seeded at low density in very low concentrations of serum (0·2 per cent) are frequently delayed for a day or two before they begin to increase in number. The rate of incorporation of [³H]thymidine is high from the very first day, however, indicating that the rate of DNA synthesis is high, but that a high rate of cell death keeps the population from increasing. Once the rapid increase in total cell number begins, however, it proceeds at a relatively rapid rate. It seems likely that the lag period is used by the cells to "condition" their own medium by releasing serum-like proteins (Rubin 1966a; Halpern and Rubin 1970).

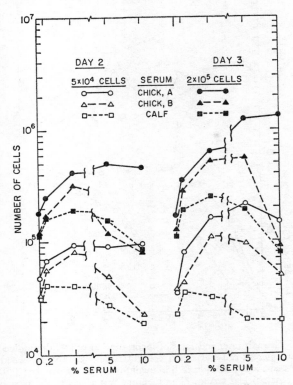

FIG. 2. Graph of various calf and chicken sera against cell concentration. Two lots of commercial chicken serum and one of calf serum were used at the concentrations shown in the abscissa to make up a medium with medium 199+2 per cent TPB. CEF were seeded at 5×10^4 and 2×10^5 cells per dish and counted on days 1 and 2.

In the presence of relatively high concentrations of serum (3·2 per cent) the cells grow rapidly at both low and high cell densities until they reach a plateau in number, at which time the rate of incorporation of [³H]thymidine decreases precipitously. Media containing high concentrations of serum obtained from such cultures are unable to support sustained growth of freshly seeded, low-density cultures of cells, and are, therefore, considered to be depleted. This is reflected in frank cell degeneration in the original culture after their maximal number is reached. There is no evidence to indicate that this restriction in number is caused by anything other than depletion of the medium.

Cells grown in lower concentrations of serum decrease in growth rate at a time when their media remain fully competent to support growth

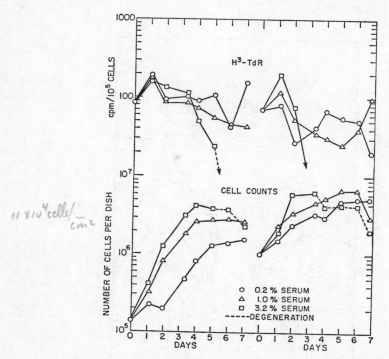

FIG. 3. Growth rates and rates of incorporation of [³H]thymidine by CEF at different cell and serum concentrations. CEF were seeded at 2 × 10⁵ and 1 × 10⁶ cells per dish in medium 199 plus 2 per cent TPB + 2 per cent chicken serum. The medium was changed the next day (= 0 days on abscissa) to media containing various concentrations of chicken serum. Cells were counted daily and the incorporation of [³H]thymidine (H³-TdR) into acid-insoluble material was determined for a one-hour period. On the fourth day the medium was harvested from some cultures in each group. Half was dialysed against the corresponding fresh medium. Dialysed, undialysed and fresh media were used to support the growth of varying numbers of freshly seeded cells to determine the degree of depletion of micro- and macronutrients. The results of the second part of the experiment are discussed in the text but not shown in the figure.

of smaller numbers of cells at a maximal rate. Several days after the rates of cell increase and [³H]thymidine incorporation have slowed down, they may increase sharply and spontaneously. One possible explanation for this is that an overgrowth-stimulating factor released by secretion or by cell death accumulates in the medium and finally exceeds a threshold concentration required to manifest its activity (Rubin 1970a, b, c). Another is that cell death decreases the population density, thereby allowing a spurt of growth in the remaining cells. No attempt has been made to distinguish between these possibilities.

The concept of "saturation density", which implies an upper limit to the number of cells in a culture, does not apply to CEF under these conditions. As long as nutrients are available in the medium even crowded cells continue to incorporate [³H]thymidine at a respectable, albeit reduced rate, and to inch up in total number. A well-defined cessation of growth accompanied by a precipitous decline in incorporation of [³H]thymidine is associated with medium depletion and is soon followed by frank degeneration of the culture (Fig. 3, 3·2 per cent serum).

However, with low concentrations of serum, the growth of the cells at high density definitely slows down before the medium loses its capacity to sustain rapid growth in freshly seeded low-density populations of cells. One cannot infer from this result that contact as such is the growth-inhibiting force. Because of their highly asymmetrical shape, a large proportion of the cells are in functional contact (i.e. exhibit electrical coupling) when only a fraction of the surface area of the dish is covered. Thus about 80 per cent of the cells are in contact when there are 5×10^5 cells in the dish, but the cell layer is not confluent until there are almost 2×10^6 cells in the dish. Despite the many contacts, cell growth continues at a high rate beyond 5×10^5 per culture.

Indeed, it is not until the cells approach or achieve confluency that a detectable change in growth rate occurs even with the lowest concentration of serum. In this region of cell density, however, there is a significant decrease in growth rate with 0·2 and 1·0 per cent serum. This "break" in the growth curves occurs at a higher cell density with 1·0 per cent than with 0·2 per cent serum. Presumably a similar break would have occurred at a higher cell density with 3·2 per cent serum had the medium not been depleted. The results indicate that if contact plays a role in inhibiting growth the effect is a quantitative not a qualitative one, and probably depends on the extent of the surface area involved in the contact. The inhibition cannot be attributed to the mere establishment of electrical coupling since this occurs as soon as contact is made (O'Lague et al. 1971).

It is noteworthy that 0·2 per cent serum, the lowest concentration which can still support growth, does so at a rate which is not less than half the maximal growth rate of the cells, which is attained with 16 times higher serum concentration. It therefore seems unlikely that the availability of serum is a precise controlling element in the regulation of growth in these cells.

At no time do we find evidence for depletion of the growth-promoting macromolecules of serum. All the media could be restored to their original growth potential, and better, at four days, by simply dialysing them against fresh medium of the same composition. Indeed T. Gurney (unpublished) has found that a medium used continuously in six passages of high concentrations of cells, but dialysed against fresh medium at each passage, retained its full growth-promoting activity. We conclude, therefore, either that the serum acts catalytically, or that the cells themselves produce a serum-like material (Halpern and Rubin 1970), which compensates for the serum used by the cells.

WOUND-HEALING EXPERIMENTS WITH CEF

"Wound-healing" experiments have a long history in the culture of CEF, and indeed occupy a whole chapter in Fischer's classic *The Biology of Tissue Cells* (1945). A recent experimental analysis of "wound-healing" of CEF in culture by Gurney (1969) shows that the growth stimulus to cells along the edge of the wound does not extend beyond one millimetre from the edge. It does not show that the cells have lost contact with their neighbours, but suggests that the presence of a free border, which allows spreading and migration, is associated with the increased growth rate. It is important to make such distinctions because an understanding of the mechanism of the population-dependent inhibition of growth will require discrimination between such phenomena as (a) increase in electrostatic surface charge (see section on pH, p. 141) which requires cell–cell proximity but not contact; (b) electrical coupling and exchange of ions, which require only limited contact, and (c) reductions in surface area and cell mobility, which depend on the extent of contact between cells (Rubin 1970a).

STIMULATION OF CELL GROWTH IN DENSE CULTURES (OVERGROWTH)

While animal sera are unique in their capacity to support growth at all cell concentrations, there are many, varied substances in addition to serum which can stimulate overgrowth in crowded cultures. Overgrowth-stimulating factor, a macromolecular substance released from Rous sarcoma cells (Rubin 1970a, b), extracts of normal cells (Fischer 1945; Rubin 1970c), hyaluronic acid (Macieira-Coelho, Hiu and Garcia-Giralt 1969), hyaluronidase, digitonin, ribonuclease, lysolecithin (Vasiliev et al. 1970) and insulin (Temin, 1967) are but a few of these and others are sure to be found. The overgrowth-stimulating effect seems neither highly specific nor very demanding. For example, I have found that medium removed from density-inhibited cultures and kept in ambient air and temperature for 6 hours is capable of stimulating a 50 per cent increase in incorporation of [³H]thymidine in 18 hours when substituted for the undisturbed medium of identical sister cultures (Table I). I have also made

TABLE I

EFFECT OF "AERATION" OF THE MEDIUM ON THE
INCORPORATION OF [³H]THYMIDINE IN DENSITY-
INHIBITED CULTURES OF CEF

Medium	Incorporation in c.p.m. per 10^5 cells at 18 hours
Undisturbed	673·5
	576·0
"Aerated"	901·0
	980·0

1×10^6 cells were seeded in medium 199+2 per cent TPB+2 per cent chicken serum. At 3 days when the cells were heavily confluent medium was removed from some cultures, allowed to stand in the room for 6 hours and replaced on some previously undisturbed cultures. 18 hours later [³H]thymidine was added for 1 hour to these cultures, and to cultures which had been left undisturbed throughout, and the incorporation into acid-insoluble material was determined.

the following paradoxical observation on growth stimulation in crowded cultures. CEF were grown for 3 days in medium 199 plus 0·2 per cent chicken serum. When the old medium was replaced by fresh medium of the same composition, there was a two to three-fold increase in both mitotic and [³H]thymidine incorporation rates within the following day (Fig. 4). Yet when the old medium was compared with fresh medium on 1-day-old cultures seeded at the same initial concentration as the original cultures had been seeded at, their relative efficacy for growth promotion was reversed; the cultures in the old medium had about twice as high a rate of mitoses and [³H]thymidine incorporation as did the cultures in the fresh medium. Thus it seems that simply upsetting the equilibrium that the 3-day-old cells had established with their medium by substituting fresh medium was sufficient to "turn on" the cells although the original medium proved superior to the fresh medium in a more recently seeded group of cultures.

These observations lend support to the proposal of Vasiliev and Gelfand (1968; Vasiliev et al. 1970) that almost any non-toxic substance added to cells in some quasi-equilibrium state with their environment is likely to speed up their growth.

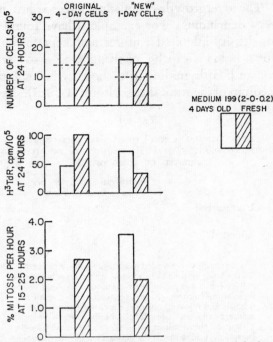

FIG. 4. Comparison of old and fresh media on old and fresh cultures.
1×10^6 cells were grown for 4 days in medium $199 + 2$ per cent TPB +
$0 \cdot 2$ per cent chicken serum. Fresh medium was of the same composition
and was substituted on some cultures. The old medium removed
from these cultures, and fresh medium, was placed on cultures seeded at
1×10^6 cells one day previously. Mitoses were accumulated with
5×10^{-8} M-Colcemid from 15 to 25 hours and counted. [³H]thymidine
incorporation was assayed from 24 to 25 hours. The total cell number
was determined at 25 hours.

INHIBITION OF CELL GROWTH BY FACTORS OTHER THAN POPULATION DENSITY

One proposed explanation for density dependent inhibition of cell
growth is that the permeability of crowded cells to components of the
medium is altered as a result of changes in the cell surface (Rubin 1970a, b;
Cunningham and Pardee 1969). I reasoned that if this were the case,
altering the concentration of components of the medium might modulate
the growth rate of the cells, and did the obvious experiments. Lowering
the concentration of substrate molecules is one approach to this problem.

I have grown cells with as little as one-fifth and even one-tenth the
usual concentration of amino acids and glucose of medium 199 to observe
the effect on growth rate. The experiment is a bit tricky because high
concentrations of cells quickly deplete such media, while very low concentrations of cells apparently leak small metabolites into the medium

and are damaged unless an adequate equilibrium can be established with components of the medium (Fig. 5). With intermediate cell concentrations, however, there is little evidence of a slowdown in growth rate with large reductions in micro-nutrients (Fig. 5). There may be some delay in

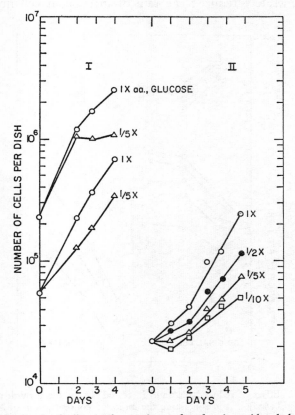

FIG. 5. Graph of cell growth at various reduced amino acid and glucose concentrations. The concentration of all the amino acids in medium 199 was reduced by diluting an amino acid mixture to the levels indicated on the graph. The concentrations of glucose were reduced proportionately. Cells were seeded at various concentrations in the appropriate amino acid, glucose mixture in medium 199 + 2 per cent TPB and 1 per cent chicken serum and counted on successive days.

initiating an increase in cell number, but once it starts it proceeds at about the same rate as in the regular medium. Nor does increasing the concentration of amino acids and glucose increase the growth rate or the cell concentration at which growth is slowed down—in fact it decreases the growth rate, the amount of these components in the regular formulation of medium 199 being just about optimal.

Another approach is to study the effects of altering the concentrations of ions in the medium. Replacing Na+ with K+ was less damaging to the cells than substituting either choline or Li+ for Na+. In agreement with the results of Cone (1969a, b), reducing the Na+ concentration by one-half moderately reduced the rate of increase in cell number, and

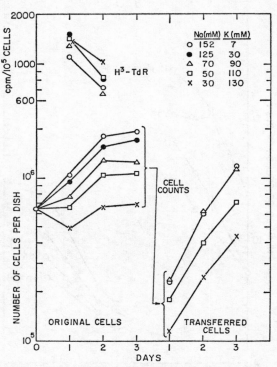

FIG. 6. Effect of reduced Na+ and increased K+ on the growth rate of CEF. Medium 199 containing the indicated concentrations of Na+ and K+, with 2 per cent TPB+2 per cent chicken serum, was used to support the growth of cells seeded the previous day at 5 × 10⁵ per dish. Cells were counted and the incorporation of [³H]thymidine estimated at daily intervals. At 3 days some of the cultures in each Na+-K+ mixture were transferred to ordinary medium 199+2 per cent TPB +2 per cent chicken serum at 2 × 10⁵ per dish, and the cells counted at daily intervals.

further reductions in Na+ had more profound effects (Fig. 6). The reduced growth rate, however, was not reflected in reduced incorporation of [³H]thymidine, as measured either by extraction of the DNA or by radioautography (Table II). (If anything the proportion of labelled cells increased as Na+ was reduced.) Transfer of the cells to regular medium after three days at various Na+ concentrations revealed some toxic effects

TABLE II

RADIOAUTOGRAPHY OF CELLS MAINTAINED IN VARIOUS CONCENTRATIONS OF Na⁺ AND K⁺

No.	Na^+ (mM)	K^+ (mM)	Total cells observed	Labelled nuclei	Percentage labelled
1	152	7	1181	56	3·1
6	70	90	559	29	5·2
7	50	110	1232	71	6·3
8	30	130	1164	75	6·5

Cells labelled for 1 hour with 1 μCi/ml of [³H]thymidine.

of low Na⁺, since only about half the cells maintained at the lowest Na⁺ concentration attached to the dish. Those that did attach grew at a normal rate. Reduction in K⁺ concentration had effects similar to those of reduced

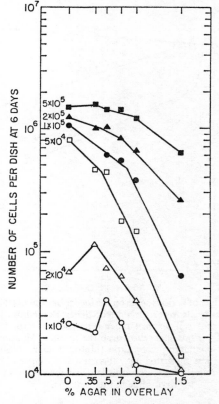

FIG. 7. Growth of various cell concentrations in a range of concentrations of Bacto-agar. Cells were seeded in medium 199 + 10 per cent TPB + 4 per cent calf serum + 1 per cent chicken serum in the concentrations shown adjacent to the curves. The next day they were overlaid with Scherer's medium + 10 per cent TPB + 4 per cent calf serum + 1 per cent chicken serum (standard for our assay of Rous sarcoma virus) with the concentrations of Bacto-Difco agar shown on the abscissa. The agar was removed at 6 days and the cells were trypsinized and counted.

Na⁺. Because of the discrepancy between the effects on cell number and on incorporation of [³H]thymidine, no clearcut statement can be made about the effect of low Na⁺ or K⁺ on growth rate. (It is probably naive in any case to suppose that altering the external ion concentration will result

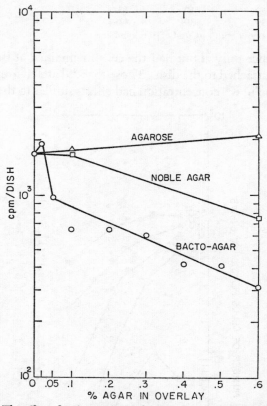

Fig. 8. The effect of various agars on the incorporation of [³H]thymidine by CEF. 1 × 10⁶ cells were overlaid with Scherer's medium + 10 per cent TPB + 4 per cent calf serum + 1 per cent chicken serum in the concentrations of various agars shown on the graph. (Agarose is a highly purified agar containing very little sulphated agar. Noble agar has about 5 times less sulphated agar than the Bacto-agar.) At 18 hours the agar was removed and the cells labelled for one hour with [³H]-thymidine in medium 199.

in proportional changes in the internal concentration of these ions, since cells have self-adjusting mechanisms for maintaining homeostasis with regard to cations; Elkinton and Danowski 1955.)

Replacing 90 per cent of the chloride ions by isethionate had no effect on either the rate of increase in cell number or the rate of incorporation of [³H]thymidine. The results of changing hydrogen and bicarbonate ions will be discussed separately below.

Bacto-Difco agar, which is added to the medium for virus assays, was found to be an effective inhibitor of growth as measured by increase in cell number (Fig. 7) or by incorporation of [³H]thymidine (Fig. 8). It was inhibitory in concentrations as low as 0·1 per cent, which is below the gelling concentration. Noble agar was less inhibitory and agarose was not inhibitory at any concentration. The inhibitory effect of agar thus appears to be related to its content of sulphated polysaccharide, and it is assumed that these strongly acidic polyanions are responsible for the inhibition. Possible mechanisms are discussed below.

EFFECTS OF PROTONS ON GROWTH

Growth of CEF in medium 199 plus 5 per cent chicken serum is most rapid between pH 7·4 and 8·2 (Fig. 9). The growth rate is markedly reduced below neutrality. As the cells grow in the presence of high concentrations of serum the pH of the medium declines continuously, because of the acid production by the cells. The pH at which cell growth slows down or stops seems to be related to the concentration of cells present. For example 1×10^5–2×10^5 cells grow at a moderate to fast rate between pH 6·65 and 6·9, but 1×10^6 cells are sharply reduced in growth rate at pH 6·95 and are stopped at pH 6·76.

The cells initially plated in low concentrations of $NaHCO_3$ buffer to achieve a low pH cause the pH of the medium to drop to a cytostatic level before they reach confluency. For example, cells initially at pH 6·9 and 7·3 have dropped to a pH of 6·5 or 6·6 by day 4, which stops cell growth at the non-confluent concentrations of $5·5 \times 10^5$ and 1×10^6 cells per dish respectively. Growth is reinitiated at a maximal rate in these cultures by simply adding $NaHCO_3$ to raise the pH to 7·6. By contrast, the cultures initiated at pH 8·2 had completely depleted the medium by 4 days, as indicated by the failure of this medium to support further growth in cells stopped by low pH (Fig. 9). There was in fact no indication that their growth had slowed down before the medium was depleted although they had grown beyond confluency.

These findings suggested that pH in combination with cell density plays a significant role in regulating cell growth. To test this possibility cells were seeded in sparse, moderate and dense populations and their capacity to incorporate [³H]thymidine was measured after the medium was changed to pH's between 7·0 and 7·7. The capacity to incorporate [³H]thymidine at the higher cell concentrations was inhibited at pH less than 7·4 on the first day after the medium change (Fig. 10). There was no indication that the lowest cell concentration was affected in this pH range. The inhibition of [³H]thymidine incorporation by pH became more pronounced on the second day, when incorporation in both of the higher concentrations of

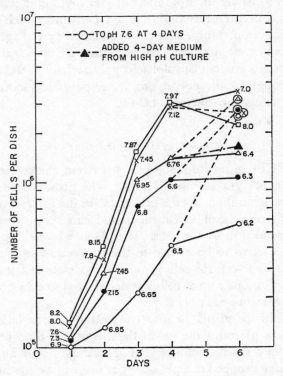

FIG. 9. Growth rates of CEF at various pH's. NaHCO₃ in varying amounts was added to NaHCO₃-free medium 199 + 5 per cent chicken serum and the medium equilibrated with the 5 per cent CO₂–air mixture of the incubator. 2 × 10⁵ cells were added to the cultures. The cells were counted daily, and the pH of the medium determined each day. On day 4, NaHCO₃ was added to some of the cultures at the lower pH's to bring them to pH 7·6, and cell counts were done 2 days later. Also, at 4 days medium was removed from the culture at highest pH, then at pH 7·97, and substituted for the medium of a culture then at pH 6·76, and the cells counted 2 days later.

cells was severely reduced below pH 7·6, while the lowest cell concentration remained unaffected even at pH 7·1. The changes in incorporation of [³H]thymidine were paralleled by changes in the rates of multiplication of the cells. There was no indication of growth inhibition at pH above 7·6 even when the cell density was equivalent to more than two monolayers. At high pH growth stops only when the medium is depleted.

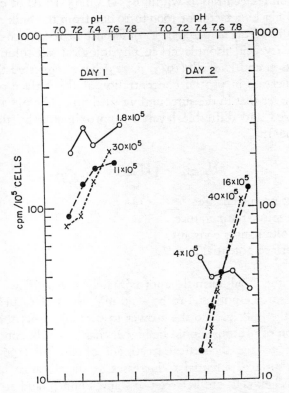

FIG. 10. Dependence of growth of cells at various concentrations on pH. Cells were seeded at 1×10^5, 5×10^5 and 20×10^5 in medium 199 + 2 per cent TPB + 1 per cent chicken serum. The next day they were switched to media at various pH's. Cell numbers, [³H]thymidine incorporation rates and the pH of the media were determined on the next 2 days. Cell numbers are not shown but are discussed in the text.

DISCUSSION

The experiments suggest that pH is a crucial factor in determining whether cell crowding leads to inhibition of growth. It is apparent that dense populations of cells are inhibited in the physiological pH range 7·0–7·4 where sparse populations show no evidence of inhibition. In effect, high population density causes an upward shift of about 0·5 pH unit in sensitivity to pH. Stated another way, it requires about three times as high a concentration of protons to inhibit a sparse population of cells as a dense population. This raises the question of whether the proton concentration at surfaces of cells is (a) different from the bulk concentration and (b) higher in crowded than in sparse cells.

The proton concentration within 0·5–1 nm (5–10 Å) of charged inter-faces may differ by a factor of 1000 to 10 000 from the bulk concentration in dilute buffers (Kavanau 1965). A difference of about 100-fold has been found for fatty acid monolayers in physiological salt solutions (Danielli 1937). Davson and Danielli (1943, p. 323) have suggested that there is a 2–10-fold increase in proton concentration at the surface of the plasma membrane over that in the surrounding medium. The pH at an interface between fixed and diffusible layers is approximated by the expression (Dawson 1968):

$$pH_{surface} = pH_{bulk} + \frac{e\psi}{2\cdot3KT}$$

where e = electronic charge
T = absolute temperature
K = Boltzmann's constant
ψ = surface potential in mV.

At 37°C, the final term is approximately $\psi/60$. If we assume the electrostatic surface potential to be −20 mV, the surface pH is 0·3 units lower than the bulk pH, so the answer to part (a) is that the cell surface concentration of protons should be higher than the bulk concentration.

According to the theoretical treatment of Gingell (1967, 1968) the electrostatic surface potential ψ increases as the distance between cells decreases. When cells are between 0·5 and 1 nm (5 and 10 Å) apart the negativity of ψ should increase by 15–40 mV (Gingell 1968). This would be expected to increase the surface proton concentration 2 to 10-fold in closely packed cells over sparsely distributed cells. Thus the answer to part (b) is yes, the concentration of protons at the surface of crowded cells should be higher than that of sparse cells, and this could account for their differing sensitivity to growth inhibition at the same bulk pH. Perhaps it is a bit early to throw another term for growth inhibition into the crowded hopper, but I suggest we keep "proton inhibition" in reserve in case the suggested mechanism is proved. Of course, if protons are indeed the controlling element in density dependent inhibition, we must ascertain whether they delay growth in the early part of the mitotic cycle or G1, and we must conceive of a mechanism to explain this effect. Sisken and Kinosita (1961) have shown that lowered pH has no effect on the length of the S and G2 phases of the cycle of kitten lung cells but does lengthen the time of telophase plus G1. This then places proton inhibition in the same part of the mitotic cycle as density dependent inhibition (Gurney 1969).

The mechanism of proton inhibition remains wholly speculative at this point. It should be pointed out, however, that Kavanau (1965) has made proton activity the cornerstone of his theory of the functional control of cell membrane configuration. He has proposed that high proton activity in the vicinity of membranes causes a change from the open or active configuration of the membrane to the closed or inactive configuration. The configurational changes are expected to affect the counterion distribution and ionic permeability of the membrane (see pp. 333–337 in Davson and Danielli 1943). The state of aggregation of the glycoproteins of the cell surface is also sensitive to pH (Gottschalk 1966) and these intercellular macromolecules are likely to be involved in growth regulation (see Burger 1971).

The proton model of cell regulation can be invoked to explain both the stimulation of cell growth by serum and the inhibition of growth by the sulphated polysaccharides of agar. Serum macromolecules are known to bind tightly to the surface of the cell, so much so that cells grown in a heterologous serum retain the antigenicity of that serum even after trypsinization, and can indeed be killed by antiserum to the heterologous serum (Hamburger, Pious and Mills 1963). F. Steck and I have found that cells grown in calf serum continue to release calf serum antigens into the medium for at least three days of successive daily medium changes with medium containing chicken serum. It seems likely that the binding of amphoteric serum proteins to the cell surface would displace many of the protons localized there, thereby increasing the surface pH of the cell and releasing it from proton inhibition. In addition, Gingell (1967) has estimated that the presence of serum molecules would reduce absolute potentials by a factor of about 0·5 because of the high dielectric constant of serum, so serum could exert its effect on local proton concentration in more than one way.

The inhibitory effect of sulphated agar can be explained in a similar way. These strongly acidic polyanions would adsorb to the cell surface, possibly by Ca^{++} bridging, much as the cell attaches to and spreads on the surface of a sulphonated polystyrene dish (Rubin 1966b). The presence of high concentrations of sulphonated groups at the cell surface would fix a high concentration of protons there. These protons would form a mobile monolayer on the polyanion surface (Kavanau 1965) and have little effect on the bulk pH of the medium. Because of the very low pK of the sulphonate group, alterations in pH would be expected to have little effect on the inhibition by agar, and I have found this to be the case.

Malignant cells are notable for their excessive production of protons, yet they are characterized by their failure to respond to density dependent

inhibition. They are also insensitive to the inhibitory effects of sulphated agar—indeed, the preferred method for producing distinctive foci of Rous sarcoma cells is by overlaying infected cultures with Bacto-agar. These observations suggest that the malignant cells are insensitive to inhibition by local pH. Appropriate conditions for testing this prediction are now being explored.

Finally, proton inhibition provides the elements of a feedback control, not only for growth but for cellular metabolism as well. Acid production is universal in metabolizing cells. If a tissue were metabolizing *in vivo* at an abnormally high rate, the local production of protons might be expected to restrain its metabolic activity in much the same manner as growth is inhibited. Of the various models for growth and overall metabolic regulation which have been proposed, this is the only one which employs in a simple feedback operation the one common product of cell activity.

SUMMARY

The growth rate of chick embryo fibroblasts (CEF) is proportional to serum concentration. At high concentrations of serum, cell growth continues at its initial high rate despite high population densities until the medium is depleted. At low concentrations of serum, the growth of the cells slows down but does not stop when the culture becomes crowded. There is no evidence that the slowing of growth is caused by depletion of the serum, and every indication that the close proximity of other cells is decisive.

A large variety of substances can stimulate "overgrowth" if added to a slowly growing confluent culture in equilibrium with its medium, including simple manipulation of the old medium of the culture.

Sulphated agar inhibits growth and thymidine incorporation in CEF, while non-sulphated agars do not. Marked decreases in Na^+ or K^+ halt the increase of the total number of cells in a population without inhibiting thymidine incorporation.

Sparsely seeded cells are reversibly inhibited at pH below 6·8. Crowded cells are inhibited at pH below 7·4. Above this pH they grow at a maximal rate until they deplete the medium. It is suggested that the concentration of protons at the surface of crowded cells is higher than that of sparse cells because of the increase in surface potential which results from the close approach of two surfaces with fixed electronegative charge. It is proposed that the increased concentration of protons at the surface of closely apposed cells stops their growth. Serum stimulation, agar inhibition and malignant transformation are discussed in terms of this model.

Acknowledgement

This investigation was supported by U.S. Public Health Service Research grant CA 05619 from the National Cancer Institute.

REFERENCES

Burger, M. M. (1971) This volume, pp. 45–63.

Cone, C. D. (1969a) In *National Aeronautics and Space Administration Scientific Publication*, L-7126.

Cone, C. D. (1969b) *Trans. N.Y. Acad. Sci.* **31**, 404–427.

Cunningham, D. D. and Pardee, A. B. (1969) *Proc. natn. Acad. Sci. U.S.A.* **64**, 1049–1056.

Danielli, J. F. (1937) *Proc. R. Soc. B* **122**, 155–174.

Davson, H. and Danielli, J. F. (1943) *The Permeability of Natural Membranes*, pp. 322–340. New York: Macmillan.

Dawson, R. M. C. (1968) In *Biological Membranes: Physical Fact and Function*, ed. Chapman, D., p. 208. New York: Academic Press.

Elkinton, J. R. and Danowski, T. S. (1955) *The Body Fluids*. Baltimore: Williams and Wilkins.

Fischer, A. (1945) *The Biology of Tissue Cells*. Copenhagen: Glydendalske Boghandel Nordisk Forlag.

Gingell, D. (1967) *J. theor. Biol.* **17**, 451–482.

Gingell, D. (1968) *J. theor. Biol.* **19**, 340–344.

Gottschalk, A. (1966) *Glycoproteins*, p. 2. Amsterdam: Elsevier.

Gurney, T. (1969) *Proc. natn. Acad. Sci. U.S.A.* **62**, 906–911.

Halpern, M. and Rubin, H. (1970) *Expl Cell Res.* **60**, 89–95.

Hamburger, R. N., Pious, D. A. and Mills, S. E. (1963) *Immunology* **6**, 439–449.

Kavanau, J. L. (1965) *Structure and Function in Biological Membranes*, vols. I and II. San Francisco: Holden-Day.

Macieira-Coelho, A., Hiu, I. J. and Garcia-Giralt, E. (1969) *Nature, Lond.* **222**, 1172.

O'Lague, P., Dalen, H., Rubin, H. and Tobias, C. (1971) *Science* **170**, 464–466.

Rein, A. and Rubin, H. (1968) *Expl Cell Res.* **49**, 666–678.

Rubin, H. (1966a) *Expl Cell Res.* **41**, 138–148.

Rubin, H. (1966b) *Publ. Hlth Rep., Wash.* **81**, 843–844.

Rubin, H. (1970a) In *II Int. Symp. on Tumour Viruses*, pp. 11–17, ed. Boiron, M. *et al.* Paris: Editions du Centre National de la Recherche Scientifique.

Rubin, H. (1970b) *Science* **167**, 1271–1272.

Rubin, H. (1970c) *Proc. natn. Acad. Sci. U.S.A.* **67**, 1256–1263.

Sisken, J. E. and Kinosita, R. (1961) *J. biophys. biochem. Cytol.* **9**, 509–518.

Temin, H. (1967) *J. cell. Physiol.* **69**, 377–384.

Vasiliev, Ju. M. and Gelfand, I. M. (1968) *Curr. Mod. Biol.* **2**, 43–55.

Vasiliev, Ju. M., Gelfand, I. M., Guelstein, V. I. and Fetisova, E. K. (1970) *J. cell. Physiol.* **75**, 305–314.

DISCUSSION

Warren: You use a sodium bicarbonate buffer for your pH experiments. This consists of two things, protons and the carbonate ions—that is, CO_2. Carbon dioxide is the product of many important reactions and one pH unit means a ten-fold difference in the concentration of hydrogen ion and theoretical CO_2 concentration. Could these large changes in CO_2 concentration account for some of your observations?

Rubin: We have examined the question of whether pH or CO_2 concentration is the critical factor by reducing the CO_2 by a factor of thirty and reducing $NaHCO_3$ proportionately to obtain the same pH values that we had with the higher CO_2 content. The results showed that pH, not CO_2 was the critical factor.

Warren: Your treatment of the bulk phase and surface pH is that of Hartley and Rowe; shouldn't you be using the ζ potential instead of ψ?

Rubin: The term used by Dawson (1968) was ψ, the surface electrostatic potential.

Warren: It would be related to electrophoretic mobility (μ).

Rubin: Electrophoresis cannot be used since it requires that the cells be individually suspended, which would destroy the relationship between cells and render the information about surface charge meaningless for the points at issue here. It might be possible, however, to use the technique of electroendosmosis.

Shodell: Did you say that when you plated 4×10^5 cells in medium adjusted to a bulk pH of $6 \cdot 6$ they were then prevented from entering even one cycle of growth?

Rubin: I have not determined the stage at which protons inhibit growth. Sisken and Kinosita (1961) found that human amnion cells were delayed in Gī by lowering the pH from $7 \cdot 8$ to $7 \cdot 1$.

Shodell: Is there any reason to think that changes in surface pH may be a function of the cell cycle itself? Could there be cyclic changes in surface pH?

Warren: There could very well be, because the surface pH should depend on fixed charge at the surface, and this appears to change during the cell cycle. There is an experiment of Mayhew (1969) who studied the electrophoretic mobility of a human osteosarcoma cell in parasynchronous growth and found that electrophoretic mobility went up during M phase. This could mean that if there is an increase in the electronegative charge of the surface, you could get an increased binding of protons. Since more protons would be present in the region of the membrane, there would be a lower effective pH at the surface of the cell in mitosis. This increased electrophoretic rate was eliminated by neuraminidase, so it was felt that it was due to an increased amount of sialic acid at the surface of the cell during M phase.

Rubin: I would think that M is the least likely stage to find the surface properties sensitive to regulatory effects, since it would appear to represent the resultant of these effects. It would be more appropriate to learn about the surface properties during G1 when the decision is made whether to pass on to S. Once this is done, M follows automatically.

Pitts: You pointed out that cell growth is dramatically changed by small changes in pH, but it should not be forgotten that other biological phenomena can be equally sensitive; enzyme activity, for instance, can be very sensitive to pH in the same sort of range.

Rubin: I agree. pH is as likely to be important in regulating metabolism as it is in growth control. The fact that acid production is a universal accompaniment of metabolic activity brings up the possibility that proton inhibition is a general feedback inhibitor of cellular activity. Protons must be of central importance to the functioning of the vertebrate organism since it has a number of devices for maintaining constant pH, such as nervous control of the respiratory exchange of CO_2, renal control of the excretion of protons, and so on.

Stoker: One difference between crowded and uncrowded cultures is that a sparsely growing cell has half its surface in contact with the substrate, whereas the crowded cell has up to twice that amount of surface in contact with substrate—that is, the dish or another cell. How does the relationship of Gingell (1967, 1968) deal with the relationship between the cell and a plastic or glass surface?

Rubin: The side of the cell facing the dish is in ridges, and therefore the area of contact with the dish is small. In spite of the apparent adhesion of the cell to the dish, most of its surface is in molecular terms far away. Furthermore there is a monolayer of serum protein at the surface of the dish under ordinary culture conditions (Rosenberg 1960).

Dulbecco: Aren't cells in tissue culture much further apart than 5 or 10 Å? Surely only in tight junctions do they reach these distances.

Rubin: This raises the old question of where the cell ends. The intercellular space may contain extracellular material which is in effect an extension of the cell, and may be important in determining the state of the plasma membrane. One attractive aspect of the theory of proton inhibition is that it states the problem of cellular interactions in quantitative terms of distance, rather than the all-or-none of contact.

Clarke: When you grow chicken fibroblasts at high pH and cell numbers reach a plateau, is further growth stimulated by fresh serum-free medium or only by serum?

Rubin: The medium is restored by dialysis against fresh medium, so the depleted elements must be small molecules. Glucose and amino acids are the likely candidates.

Montagnier: If one works with a very low concentration of cells, say

10^3 or 10^4 cells per 60 mm plate, one finds the reverse of your pH effect, namely that high pH becomes inhibitory.

Rubin: We find that prolonged exposure to pH greater than 7·6 is harmful to cells, especially when their concentration is low. On a short-term basis, however, DNA synthesis in crowded cultures is stimulated by pH values up to 8·2 and even higher, but very high pH will eventually damage even the crowded cells.

Montagnier: Could one conclude from your theoretical calculation that the shape of the cells is important for ψ, and that for a spherical shape the charge density should be increased?

Rubin: Since the spherical cell has the minimum surface-to-volume ratio, it is not impossible that the surface charge density is higher than in the cell spread at the surface of the dish, and this might play a role in the requirement that cells spread before they can grow in culture.

Bard: You said that you have no evidence that your cells ever really stop growing. What is the maximum number of cells you have counted on a 5 cm dish?

Rubin: The maximum number counted was 17×10^6; at this point the sheet tends to retract, and the cells come off the dish. But we had the equivalent of about eight or nine monolayers. That doesn't mean that there are necessarily eight or nine layers of cells over the whole dish. The cells get squeezed into smaller and smaller areas, but there are plenty of patches of two or more layers. There is of course a tendency for fibroblasts in culture to form monolayers but it is not terribly strong and when you look carefully there is tremendous cytoplasmic overlapping, even in a so-called monolayer.

Stoker: Dr Rubin, can you predict from your pH theory which is the cell which grows best, in other words, which has the lowest value of ψ?

Rubin: You cannot predict the ability of the cell to grow from its surface potential in the proton-inhibition model. This would depend on the sensitivity of whatever unit in the cell responds to the local proton concentration. It would appear however, that malignant cells are less sensitive to local proton concentration than are normal cells (Rubin, unpublished).

Stoker: When would you expect the lowest concentration of protons?

Rubin: When the cells are most spread out and most isolated from other cells. You would then have the lowest surface charge density, the greatest interaction with serum molecules and the least effect of other cells in increasing the surface potential. Also, you would have a lower local concentration of acidic molecules produced by the cells themselves. This might well include acid mucopolysaccharides, which probably polymerize at the cell surface only in dense cultures where their concentration is high.

REFERENCES

DAWSON, R. M. C. (1968) In *Biological Membranes: Physical Fact and Function*, pp. 203–232, ed. Chapman, D. New York: Academic Press.

GINGELL, D. (1967) *J. theor. Biol.* **17,** 451–482.

GINGELL, D. (1968) *J. theor. Biol.* **19,** 340–344.

MAYHEW, E. (1969) *J. gen. Physiol.* **49,** 717.

ROSENBERG, M. D. (1960) *Biophys. J.* **1,** 137–159.

SISKEN, J. E. and KINOSITA, R. (1961) *J. biophys. biochem. Cytol.* **9,** 509–518.

SV40 TRANSFORMATION AND CELLULAR GROWTH CONTROL

G. J. TODARO, C. D. SCHER AND H. S. SMITH

Viral Leukemia & Lymphoma Branch, National Cancer Institute, Bethesda, Maryland, and Meloy Laboratories, Springfield, Virginia

WE have been concerned with the growth properties of cells in tissue culture and the changes induced in these cells which may mimic changes to the neoplastic state *in vivo*. With this aim in mind, we set out several years ago to establish permanent mouse embryo lines under a variety of different culture conditions (Todaro and Green 1963). The original primary or secondary cultures of mouse embryo cells were found to grow poorly when plated either very sparsely or under very crowded culture conditions, but were able to grow rapidly when plated at moderate cell densities (approximately 10^5–10^6 cells per 20 cm^2 Petri dish). This difference in growth rate as a function of cell density can be seen in Fig. 1. For comparison, we have included the growth properties of L cells, a mouse cell line that had been in culture for many years (Earle 1943). The latter cells show the same overall doubling time whether they are grown under sparse, moderate or crowded culture conditions. By transferring the embryo cells every three days under relatively sparse culture conditions (3×10^5 cells per dish) the 3T3 cell line was established (see Table I). 3T3 cells differ from the original mouse embryo

TABLE I

CELL LINES DERIVED FROM MOUSE EMBRYO CULTURES

Cell line	Interval between transfers (days)	Number of cells plated at each transfer ($\times 10^{-5}$)
3T3	3	3
3T6	3	6
3T12	3	12
6T6	6	6
6T12	6	12

cells in several ways. The cells have gained the ability to grow under sparse culture conditions but they cease dividing at a lower cell density than the original embryo cells. In contrast, when we established the original mouse embryo culture by transferring the cells every three days at 12×10^5 cells per dish, we established a line, 3T12, which is unable to grow at low cell densities but has gained the ability to grow to high saturation densities (Todaro and Green 1963).

When the same culture conditions were used to develop cell lines from embryo cells of an inbred strain of mouse, Balb/c, essentially the same cell culture properties were obtained (Aaronson and Todaro 1968*a*).

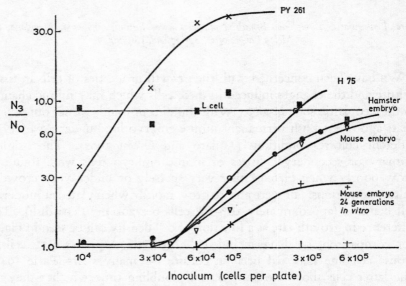

FIG. 1. Doubling time of the cell population during a three-day interval when cells are seeded at differing numbers of cells per plate. H 75, a hamster tumour cell line produced *in vivo* by polyoma virus. PY 261, a hamster cell line transformed in tissue culture by polyoma virus.
L cells, mouse cell line L 929.

The Balb/c 3T3 lines and Balb/c 3T12 lines were then tested for their ability to cause tumours when injected into weanling irradiated Balb/c mice. We found that animals inoculated with Balb/c 3T12 cells readily formed tumours, whereas Balb/c 3T3 cells under similar conditions did not form tumours (Aaronson and Todaro 1968*b*). This indicated that the property of growth to high saturation densities was related to tumorigenicity and that it would be fruitful to study changes induced in 3T3 cells which would enable them to grow to high saturation densities.

3T3 or Balb/c 3T3 cells can be altered so as to be able to grow to high saturation densities by a number of methods. The cells can be transformed by oncogenic DNA-containing viruses such as SV40 and polyoma (Todaro, Green and Goldberg 1964) and by oncogenic RNA-containing viruses such as the murine sarcoma virus (MSV) (Todaro and Aaronson 1969) and the avian sarcoma viruses (unpublished experiments). Balb/c 3T3 cells surviving irradiation with ultraviolet light or X-rays exhibit the transformed property of growth to high saturation densities (Pollock, Aaronson and Todaro 1970). Transferring the cells, even cloned populations, under conditions that selectively favour the growth of cells that can grow to higher saturation densities leads to the appearance of "spontaneous" transformants.

When 3T3 cells are transformed so that they grow to high saturation densities, they gain other growth properties as well. The Balb/c transformants gain the ability to form tumours when injected into weanling irradiated Balb/c mice. Virally transformed 3T3 and Balb/c 3T3 cells gain the ability to grow in agar suspension (Black 1966) and to grow in medium lacking serum factor(s) necessary for the growth of 3T3 (Holley and Kiernan 1968; Jainchill and Todaro 1970). For the past year, we have been particularly concerned with this last characteristic, namely the ability to grow in medium depleted of serum factor(s) essential for the growth of 3T3 and Balb/c 3T3.

This "factor-free" medium was prepared using Dulbecco's modification of Eagle's medium plus 20 per cent agamma newborn calf serum, made gamma-globulin free by a modified Cohn alcohol precipitation (North American Biological, Inc.). The medium was further depleted on confluent Balb/c 3T3 cells to remove any residual growth-promoting activity. After depletion, the medium was diluted with an equal volume of serum-free medium (the final serum concentration was then 10 per cent) to assure that the medium was not depleted of essential nutrients. A more direct way of preparing growth-factor-free serum involves heating prepared agamma calf serum to 70°C for 30 minutes. The heating destroys any residual ability of the serum to support division of sparse cultures of Balb/c 3T3, whereas those serum factors needed for cell survival remain active. Heating to 56° or 60° C has little effect on residual growth-stimulating activity. This heat treatment avoids the need to further deplete the medium on cells and thus rules out the possibility that the selective nature of the medium is due to inhibitors produced by confluent cells (Smith, Scher and Todaro 1971).

The "factor-free" medium was used to ask some questions about the early interactions between SV40 and Balb/c 3T3 cells. A particular clone (A31) with a maximum cell density of 1.0×10^6 cells per dish was used in all experiments. Early work had shown that four to eight cell divisions after

SV40 infection were necessary for phenotypic expression of the loss of density dependent inhibition (Todaro and Green 1966). We wondered whether the cells would gain the ability to grow in "factor-free" medium immediately after infection. To test this, cells were plated sparsely in "factor-free" medium. Four days after the cells had been plated, some of

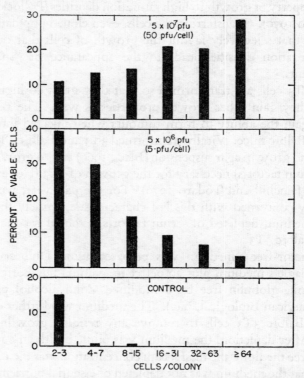

FIG. 2. Effect of infection with SV40 on colony size in factor-free medium. Cells at confluence and in factor-free medium were exposed to SV40 at a multiplicity of 5 and 50 pfu/cell and then inoculated into factor-free medium at 5000 cells per plate. The number of cells per colony was determined every few days. The values shown are at 11 days after SV40 infection.

the plates were infected with SV40. The virus induced many cells to synthesize DNA (as measured by radioautography) and these cells went on to divide several times. Some of the colonies contained more than 100 cells. To determine whether a functioning virus genome was necessary to induce cell division in "factor-free" medium, the virus was inactivated by ultraviolet light. A dose of ultraviolet light that caused a 200-fold loss of infectivity on green monkey kidney cells (pfu/ml) caused an approximately 10-fold loss of transforming efficiency in Balb/c 3T3 cells and decreased the ability of the virus to induce the growth of A31 colonies in factor-free medium by at least a factor of 10. This suggested that the viral genome rather than the viral protein (coat or internal) had the capacity to induce cell division. We also studied the induction of cell division by SV40 at

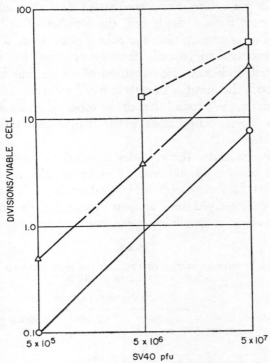

FIG. 3. The effect of virus inoculum on the number of cell divisions induced in growth factor-free medium. Cultures were infected with 0·5 ml of virus suspension; cell divisions were scored 10–11 days after infection. Three separate experiments are shown. In each, control values from mock-infected cultures were subtracted from the total cell divisions.

□, confluent cultures (1 × 10⁶ cells) that had not had their medium changed for three days were infected with SV40 and plated on factor-free medium; control cultures had 0·3 divisions per viable cell. △, confluent cultures (1 × 10⁶ cells) that had not had their medium changed for three days were infected with SV40 and plated in factor-free medium; control cultures had 0·7 divisions per viable cell. ○, sparse cultures (1 × 10² cells) already in factor-free medium were infected with SV40; control cultures had 0·02 divisions per viable cell.

low virus multiplicities. We found that nearly the same number of cells was induced to divide at an input multiplicity of 5 pfu/cell as at an input multiplicity of 50 pfu/cell; in the latter case, however, the average colony size was considerably larger. At both multiplicities, colonies were observed that contained more than 100 cells (Fig. 2).

The transformation frequency of Balb/c 3T3 is known to be directly proportional to the multiplicity of virus used to infect the culture. To determine whether the total number of cell divisions induced was also proportional to the multiplicity, the number of colonies as well as the number of cells per colony was scored 10–11 days after infection. Since each colony arose from a single cell, the number of cell divisions per colony was taken as one less than the colony size. Thus, a colony which has grown from one to 16 cells (four generations) has undergone 15 independent cell divisions. This method of expressing the data takes into account both the number of cells induced to divide and the number of cells per colony. We found that the number of divisions induced per viable cell was directly proportional to the virus inoculum (Fig. 3) (Smith, Scher and Todaro 1971).

To determine whether the colonies induced to grow in factor-free medium were positive for the viral T antigen, cells were plated relatively sparsely and infected with 10^8 pfu/ml of SV40. Twelve days later the coverslips were assayed for T antigen by indirect immunofluorescence (Table II). No colonies containing more than four cells were found in the

TABLE II

SV40 T ANTIGEN STATUS OF CELLS IN COLONIES INDUCED TO
GROW BY EXPOSURE TO SV40

	Number of colonies		
Cells per colony	Negative*	Mixed**	Positive†
5–20	20	1	1
21–50	8	4	3
> 50	8	9	3

* All the cells in the colony negative for the SV40 T antigen.
** Both positive and negative cells in the same colony.
† Every cell in the colony positive for the SV40 T antigen.

mock-infected control. We found that approximately 60 per cent of the colonies induced to divide by SV40 were negative for T antigen; these included colonies that had grown to over 50 cells. In other colonies every cell in the colony was positive for T antigen. One of these positive colonies contained only nine cells. In still other colonies, some of the cells were positive for T antigen while other cells in the same colony were clearly negative (Smith, Scher and Todaro 1971). These observations indicated that many of the cells induced to divide in factor-free medium no longer had demonstrable levels of at least one virally induced protein, the SV40 T antigen.

The ability to grow to a high saturation density is also characteristic of cells transformed by SV40. We found that only 10–15 per cent of the colonies that were induced to grow in factor-free medium lost density dependent inhibition. The experiment was done by noting the location on the dish of all the colonies present 11 days after infection; the cells were then shifted to complete medium and allowed to grow until they reached confluence. The plates were then stained and scored as in a standard transformation assay (Todaro 1969). Thus only 10–20 per cent of the colonies induced by SV40 to grow in factor-free medium lost contact inhibition of cell division or were positive for T antigen. This result would be expected if SV40 only temporarily changed the growth properties of the cells. Alternatively, the selection procedure may have allowed us to recognize transformed cells which had permanently associated SV40 but nevertheless were still subject to contact inhibition. To distinguish between these two possibilities, 16 colonies that had grown in factor-free medium after SV40 infection were isolated and grown up to mass cultures. When their progeny were tested for properties associated with permanent viral transformation, it was clear that both situations existed.

Several different classes of colonies were found (Table III). The first

TABLE III

GROWTH PROPERTIES OF CLONES ISOLATED BY THEIR ABILITY TO GROW IN FACTOR-FREE MEDIUM AFTER EXPOSURE TO SV40

Class	Number of clones	Maximum cell density ($\times 10^6$/plate)	Cloning efficiency in factor-free medium (percentage of control)	Persistence of T antigen
I	(6)	6–25	5–70	+
II	(8)	0·8–1·6	<0·1–<1·0	−
III	(2)	0·9–1·2	30–50	+
Control	–	0·8–1·2	<0·1–<0·5	−

class (six of the 16 clones) had the properties associated with the transformed state. They all had SV40 T antigen and their saturation densities ranged from 6·0 to 12·5 × 10^6 cells per plate. The colonies in this class all retained the ability to grow in factor-free medium. Their plating efficiency in this medium ranged from 5 to 64 per cent of the plating efficiency in complete medium, values that are comparable to those found for SV40 transformants obtained using the standard transformation method. The second class (eight of the 16 clones) had none of the properties of transformed cells; these are presumed to be the abortively transformed cells (Stoker 1968; Stoker and Dulbecco 1969). Their saturation densities in complete medium were similar to normal Balb/c 3T3 cells. They did not grow in factor-free medium and did not show any persistence of the SV40 T antigen. None of the colonies tested yielded virus after fusion with permissive cells (Smith, Scher and Todaro 1971).

6*

Two of the 16 colonies did not fall into either class. They grew to the same saturation density as normal cells, but were still able to grow in factor-free medium, though more slowly than the other transformants. They also had SV40 T antigen, and, after fusion with monkey kidney cells, yielded SV40. These transformed colonies (class III) would not have been detected by the standard transformation assay. Another colony, 10 A-4, which has not yet been placed in any of the three classes, was able to grow slowly in the factor-free medium, although it had the same saturation density as normal Balb/c 3T3 cells. In this clone, however, no additional evidence of the persistence of viral information has yet been obtained; it was negative for T antigen, and did not yield virus after fusion. More direct tests for the presence of viral information in this clone are in progress using DNA-RNA hybridization techniques.

In summary, then, we found that the selective property of factor-free medium offered the opportunity to study cellular growth changes induced by SV40 early after infection. Cells in factor-free medium are induced to divide after infection with SV40 and colonies containing more than 100 cells are formed. After infection many cells transiently gain the ability to grow in factor-free medium without becoming permanently trans-formed by the virus. In addition, growth in factor-free medium allows the recognition of transformants that remain contact inhibited and, therefore, would not be recognized by the standard transformation assay.

These experiments raise the question of whether the viral genome is physically lost from the cells that no longer have any evidence of viral function or whether the genome can remain present but is "switched off" by the host cell.

We then turned our attention to the question of whether each kind of transformation (SV40, murine sarcoma virus and "spontaneous") alters the cell in the same way. We asked whether all transformed cells were also able to grow in factor-free medium. Previous work had shown that all virally transformed cells, whether transformed with RNA or DNA viruses, are able to grow in medium lacking serum factor(s) (Jainchill and Todaro 1970). However, more recent studies have shown that many "spontaneous" transformants grew poorly if at all in this medium. These "spontaneous" transformants grew to equally high saturation densities as the virally transformed lines (Smith and Scher 1971). Thus, growth to high saturation densities is not always the result of a lowered requirement for serum factor(s) (Holley and Kiernan 1968; Clarke et al. 1970).

Even with viral transformation, there are two ways that the property of loss of "contact inhibition" and the property of growth in the absence of serum factor(s) can be dissociated. First, immediately after infection with SV40, Balb/c 3T3 cells gain the ability to grow in factor-free medium; however, 6–8 generations are necessary before the cells lose "contact inhibition". Secondly, there are SV40-transformed lines, as described above, that have only one of the growth properties associated with viral transformation, the ability to grow in factor-free medium. The simplest hypothesis consistent with all the data is that the two growth properties are independent changes and that oncogenic viruses usually induce both changes.

Balb/c 3T3 cells transformed by murine sarcoma virus have acquired the ability to grow in factor-free medium. However, if the same experiment illustrated in Fig. 2 for SV40 infection is done using MSV, no induction of cellular DNA synthesis or cell division is found. If the cells are maintained in complete medium for 5 to 7 days and then shifted to factor-free medium they are able to continue dividing. MSV, unlike SV40, then, is unable, *itself*, to induce cellular DNA synthesis. However, after four to six cell divisions the phenotypic expression of transformation—the ability to grow in factor-free medium—has occurred. SV40 can induce even the initial round of cell division in factor-free medium, indicating at least one difference in the mode of action of the two types of oncogenic viruses. Another difference is in the nature of the virally induced alterations in the cell membrane. MSV, but not SV40, changes the kinetics of glucose transport in Balb/c 3T3 cells, resulting in a much more efficient uptake of glucose in cells transformed by the oncogenic RNA viruses (Hatanaka, Todaro and Gilden 1970).

Some of the conclusions that we have drawn from these experiments are pertinent to the problem of the development of cancer in the animal. Growth in factor-free medium is a different growth property from the loss of contact inhibition. All virally transformed cells so far tested are able to grow in factor-free medium whereas many spontaneous transformants grow poorly, if at all. In some cases, such as with SV40 transformation, the cells gain this transformed property before other transformed properties, like the loss of contact inhibition, are phenotypically expressed. Thus, growth in this medium offers a new transformed property with which one can test for potential tumour viruses. When permanent transformation is dissociated from abortive transformation by allowing "fixation" (Todaro and Green 1966) (3–4 cell divisions in complete medium before cloning in factor-free medium) we find that 10–20 per cent of the transformants have no higher saturation density than does Balb/c 3T3 itself. These transformants clearly would not be recognized in the standard transformation assay.

We have tested the feasibility of this approach by studying the growth properties in factor-free medium of a normal human fibroblast and the cells from this line that have been transformed by SV40. Normal human cells, like 3T3 cells, are unable to grow in this medium, whereas transformed human cells gain the ability to grow in factor-free medium (Fig. 4).

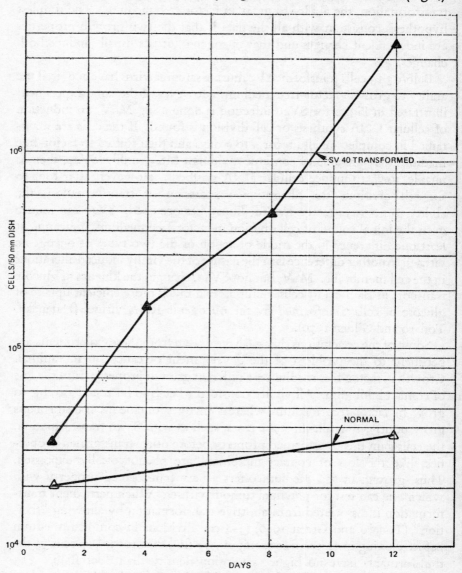

FIG. 4. Human cells in factor-free medium. Cells per Petri dish (50 mm diameter) were counted after one day and again after 12 days of growth in factor-free medium. △, human diploid cell strain L.S.; ▲, SV40-transformed L.S. cells (clone 65).

Methods have been described that favour the selective growth of tumour cells of animal and man. The ability to grow in agar is one property of tumour cells that normal cells lack (McAllister and Reed 1968). The ability to grow on the top of a confluent monolayer of normal cells is another property possessed by some but not all tumour cells (Aaronson, Todaro and Freeman 1970). Whether "factor-free" medium can be used to favour selectively the growth of tumour cells freshly explanted from the body is not yet known.

SUMMARY

Abortive transformation of Balb/c 3T3 cells by SV40 virus has been studied using a selective medium lacking serum growth factor(s). This factor-free medium supports the growth of SV40-transformed but not normal Balb/c 3T3 cells. After infection with SV40, cells were induced to synthesize DNA and divide. The cells were plated sparsely and the growth of individual colonies was followed microscopically. The number and size of the colonies scored 10 days after infection depended on the virus multiplicity. A functional viral genome was necessary since UV-irradiation inactivated the ability of SV40 to induce the formation of cell colonies. The data suggest that SV40 can transiently alter the growth properties of Balb/c 3T3 cells without becoming permanently associated with the host cell. Cells induced to divide in factor-free medium were shifted to standard medium and grown to confluence. Most of the colonies did not grow to high saturation density. Ten days after infection many colonies induced to divide in factor-free medium had no cells with T antigen. These colonies contained as many as 100 cells. In others, all cells had T antigen. Many colonies isolated from factor-free medium stopped dividing at low saturation density, no longer grew in factor-free medium, were negative for T antigen, and did not yield infectious virus upon fusion with permissive cells. Several SV40-transformed clones have been isolated that grew in factor-free medium but stopped dividing at low saturation density. These cloned lines represent new transformants that would not be detected in a standard transformation assay. Factor-free medium has been useful in the study of early virus–cell interactions as well as being an independent method of selecting virally transformed cells.

REFERENCES

AARONSON, S. A. and TODARO, G. J. (1968a) *J. cell. Physiol.* **72**, 141–148.
AARONSON, S. A. and TODARO, G. J. (1968b) *Science* **162**, 1024–1026.
AARONSON, S. A., TODARO, G. J. and FREEMAN, A. E. (1970) *Expl Cell Res.* **61**, 1–5.

BLACK, P. H. (1966) *Virology* 28, 760–763.

CLARKE, G. D., STOKER, M. G. P., LUDLOW, A. and THORNTON, M. (1970) *Nature, Lond.* 227, 798–801.

EARLE, W. R. (1943) *J. natn. Cancer Inst.* 4, 165–212.

HATANAKA, M., TODARO, G. J. and GILDEN, R. V. (1970) *Int. J. Cancer* 5, 224–228.

HOLLEY, R. W. and KIERNAN, J. A. (1968) *Proc. natn. Acad. Sci. U.S.A.* 60, 300–304.

JAINCHILL, J. and TODARO, G. J. (1970) *Expl Cell Res.* 59, 137–146.

McALLISTER, R. M. and REED, G. (1968) *Pediat. Res.* 2, 356–360.

POLLOCK, E. J., AARONSON, S. A. and TODARO, G. J. (1970) *Int. J. Radiat. Biol.* 17, 97–100.

SMITH, H. S. and SCHER, C. D. (1971) *Nature, Lond.* in press.

SMITH, H. S., SCHER, C. D. and TODARO, G. J. (1971) *Virology* in press.

STOKER, M. (1968) *Nature, Lond.* 218, 234.

STOKER, M. and DULBECCO, R. (1969) *Nature, Lond.* 223, 397.

TODARO, G. J. (1969) In *Fundamental Techniques in Virology,* pp. 220–225, ed. Habel, K. and Saltzman, N. P. New York: Academic Press.

TODARO, G. J. and AARONSON, S. A. (1969) *Virology* 38, 174–202.

TODARO, G. J. and GREEN, H. (1963) *J. Cell Biol.* 17, 299–313.

TODARO, G. J. and GREEN, H. (1966) *Proc. natn. Acad. Sci. U.S.A.* 55, 302–308.

TODARO, G. J., GREEN, H. and GOLDBERG, B. D. (1964) *Proc. natn. Acad. Sci. U.S.A.* 51, 66–72.

DISCUSSION

Macpherson: Your class III colonies obtained after SV40 infection had normal saturation densities and were resistant to superinfection with SV40. Do you know if they are also resistant to murine sarcoma virus or other transforming viruses? I was wondering whether resistance is a cellular property.

Todaro: We haven't tested that. I should mention that in the same assay in depleted medium one cannot stimulate cells to divide with high-titred murine sarcoma virus. For MSV transformants, one has to infect the cells and allow them to go through one to three (we are not yet sure how many) cell divisions before plating them in the depleted medium; then they are able to grow and continue to grow indefinitely. But MSV seems to differ from SV40 in that it seems to be unable to initiate the first round itself, although the fully transformed cells have this property. The failure of MSV-infected cells to resume cell division is perhaps surprising, since these viruses carry in with them both RNA-dependent DNA polymerase activity and a DNA-dependent DNA polymerase.

Pitts: The term "agamma serum" is being widely used now, but is it a good term? Foetal calf serum is free of gamma globulin but contains growth factors.

Stoker: We call it gamma-depleted serum.

Todaro: The saturated ammonium sulphate used in its preparation would bring down a lot of things. However, we have been more interested in getting a serum that would have the properties that we want, namely to allow 3T3 cells to stay healthy but not to divide. This seems to work (Jainchill and Todaro 1970).

Rubin: Fifty per cent ammonium sulphate should salt out the euglobulins including β-globulin, as well as the gamma globulins. It probably would even get some of the albumin.

Holley: There will still be 15–20 per cent of the serum protein left.

Taylor-Papadimitriou: But is it not in the β-globulin fraction that you find the survival factor?

Holley: We would not attribute survival activity to any known serum fraction.

Taylor-Papadimitriou: We have been using a similar system in studying the effect of interferon on the abortive transformation of BHK21 cells by polyoma virus (Taylor-Papadimitriou and Stoker 1971). The virus stimulates the cells to synthesize DNA in the complete absence of serum, but after a while the cells deteriorate. In order to follow mitoses and changes in the pattern of orientation of the cells, we included gamma-depleted serum in the medium to a final concentration of 0·8 per cent, since in this medium BHK21 cells were known to survive well. The DNA synthesis induced by the virus could be shown to be followed by a round of mitoses but again the cells did not survive well even in the gamma-depleted serum. The abortively transformed BHK21 cells therefore, while able to synthesize DNA in the absence of serum, show a higher requirement for serum factors concerned with survival than do the original BHK cells and I wonder whether the ability to grow in gamma-depleted serum is limited to transformed lines of 3T3 cells. When you say your cells "grow" in the serum, do mass cultures of, say, 3T12 grow well and have you any information on polyoma-transformed 3T3 cells?

Todaro: Both the 3T12 cells and the transformed 3T3 cells, including the doubly-transformed ones, grow well in gamma-depleted serum. They will continue to divide indefinitely in it, albeit at a slower rate. Have you tried your agamma serum on 3T3 cells?

Taylor-Papadimitriou: No, we haven't.

Bürk: We did this; 3T3 cells didn't grow and SV3T3 did grow.

Todaro: So it's the cells, not the serum.

Dulbecco: Dr Todaro, you mentioned that it is possible to obtain different kinds of transformed cells, depending on the selective conditions. Do you imply that these transformed cells may not ultimately be identical?

Todaro: No, I think they are quite different.

Dulbecco: Are they the results of a long selection after viral infection?

Todaro: The cells are picked as soon as colonies become visible, then recloned and tested. Those selected for ability to grow in gamma-depleted serum are tested fairly soon. Of the cells transformed either by SV40 or by RNA viruses or spontaneously, some were recently obtained and others have been in long-term culture.

Dulbecco: What are the frequencies with which the various types are obtained; do they differ?

Todaro: Of the first 16 clones tested in gamma-depleted medium, two had properties that would not have been picked up in another assay. Similarly, of eight clones that were "spontaneously" transformed derivatives of clone A31 of the Balb/c 3T3 cell line, three could grow in the depleted medium, although they had never been tested before, but five others did not grow or grew poorly. So these two properties (growth to high saturation density and growth in serum factor-limited medium) don't seem to go together. You could interpret this in either of two ways. The first is to say that we are selecting for one property and are finding that a high fraction, a majority, in fact, also have the other property that was not selected for; or, alternatively one can focus on the exceptions where a particular clone acquires one property but not the other, demonstrating that the two properties can be separated from each other (Pollack, Green and Todaro 1968).

Dulbecco: The important question for transformation is whether you get different cells because something is different in the way transformation occurs or whether the differences are accidental. I suggest this because your class III cells are essentially like Pollack's revertants (Pollack, Green and Todaro 1968).

Todaro: But his were obtained by secondarily imposing another selective pressure. Here there is no selection one way or the other for the second property.

Dulbecco: However, if you think that Pollack's revertants are due to a cellular mutation which gives this particular property, the same cellular mutation could already exist in untransformed 3T3 cells and may just be revealed by transformation.

Todaro: Yes, but one point is that we may be underestimating the transformation frequency in any one test; we may not by any single test be detecting the whole range of transformants.

Stoker: The lack of super-transformability could be explained by it being a cellular mutation.

Dulbecco: It would be worthwhile doing an experiment to test whether a cellular mutation is involved.

Todaro: An argument that it is, in fact, a cellular mutant is that the virus that comes out is normal in its transforming ability.

Dulbecco: This could be tested, because if there are two out of 16 clones with this mutation, and the mutations pre-exist in the 3T3 population, these mutants occur with a high frequency; then, if you pick 20 clones of 3T3 cells and transform them you might find one or two which only give rise to transformed clones of class III.

Macpherson: How stable are the class III cells? Do variants arise that grow to high density?

Todaro: This has not been tried but I suspect we would find this, as we can from any 3T3 population get variants that will grow to very high cell densities.

Macpherson: It would be interesting to determine the frequency with which variants of this type are formed.

Clarke: Do class III cells grow in agar?

Todaro: We haven't tested how these two transformation assays relate to the agar assay.

Clarke: May I ask two questions which may have been answered in work of which I am ignorant. Firstly, when one transforms mouse embryo cells with virus do the cells acquire, without subsequent selection, a high cloning efficiency? Secondly, in the process of establishing permanent mouse lines, are the two properties of growth at high density and high cloning efficiency entirely independent?

Todaro: This would be difficult to do in primary mouse embryo cells, because one of their characteristics is that they are not very easy to clone; that is one of the major advantages of using continuous lines. They allow you to do this kind of experiment.

Pitts: You can select a 3T12 type from a 3T3 population; can you do the reverse?

Todaro: Yes. One property can be selected independently of the other and once you have one property you can get the other, either way.

Pitts: Is this simply a selection of cell variants?

Todaro: No; 3T12 can be adapted. As long as you ask the cell to do something which is within its capacity to do, and you have some patience, it will do it.

Pitts: But culture at low density does not select for cells which have a low saturation density.

Todaro: You can't do it all at once. If we want a 3T12 cell which cannot grow when inoculated at less than 10^5 cells per plate to grow at low density, we would not start by plating a hundred cells. We would plate 10^4 cells; some would survive and grow. These would be plated at 10^3 cells per plate and so on.

Pitts: And would those cells stop growing at a low saturation density like 3T3? That is the question.

Todaro: No; they are still 3T12 cells that can now also grow from low cell densities.

Pontecorvo: Is it more difficult to do the back selection than the original one, or about the same? If you start from the embryo and put some cells in 3T3 conditions and some in 3T12 conditions, then take the 3T3 cells once they are established and try to select the other type out, is it easier, more difficult or about the same?

Todaro: Once you establish a cell line—that is, once it has become aneuploid—it is fairly easy to select variants with properties that you want. The L cell, for example, has probably been through so many different kinds of selection pressures in its history that it can survive almost anything. This is the pattern of 3T3. It is not difficult to get variants but not all in one step. However, it is easier than with the diploid cells. Aneuploid cells have this variability, this ability to adapt to new circumstances that diploid cell cultures lack.

Stoker: By adapt, you mean selection of variants?

Todaro: Yes; most likely it is selection of pre-existing variants.

Rubin: We need a clearer definition of what is meant here by selection. Are you selecting a variant which pre-exists in the embryo, or are you creating new types by subjecting the cells to culture and then selecting the most suitable type? The evolution of the 3T3 type of cell includes many successive passages and includes a prolonged period of decline followed by the appearance of a rapidly growing, highly aneuploid line of cells (Todaro and Green 1963). If the 3T3 type of cell had been present from the beginning it would have quickly outgrown the other cells during the decline period. It seems more likely that a new type appears in culture. Mitotic abnormalities are common in monolayer cultures, and these could lead to a wide diversity of chromosome complements among the cultured cells. It is from this diversity that the final selection would be made. Each independent isolation of a 3T3 line might then be genetically unique, and not strictly comparable to other lines in its properties. Furthermore, variation and selection probably are continuous, so that any given line is changing its properties all the time. When one uses cells fresh from the embryo, one avoids these uncertainties.

Pontecorvo: The second selection should be more difficult than the original one because you have restricted the variation in your cell population.

Rubin: The second time around is more efficient, however, since you don't have to go through the protracted period of declining growth rate.

Weiss: It is not clear to me why growing at low cell density selects against the ability to grow at high cell density. I don't see why a 3T3 cell should have a much lower saturation density than a primary mouse embryonic fibroblast.

Todaro: I'm not really sure either! The saturation density is a million for 3T3 and the primary embryo cells would grow to 5 or 6 million. I think maybe it's because of the conditions involved; the cells have always been in serum-factor excess. They are given a lot of serum factor per cell. The medium is changed regularly. We are presumably also selecting for cells that attach well to plastic, inadvertently, because we transfer the cells, give them time to attach and then wash off the unattached cells. So we must be selecting cells that attach more quickly. I'm not certain why it should turn out this way, but it does, not only with the original 3T3 line but also with the line from Balb/c embryos and the line derived from NIH Swiss embryos.

Weiss: Did you try zig-zag experiments, growing the cells alternately at very low and very high densities, to select for both ability to clone and ability to grow at high density?

Todaro: Yes. In fact, most continuous lines have that possibility.

REFERENCES

JAINCHILL, J. and TODARO, G. J. (1970) *Expl Cell Res.* **59,** 137.
POLLACK, R., GREEN, H. and TODARO, G. J. (1968) *Proc. natn. Acad. Sci. U.S.A.* **60,** 126.
TAYLOR-PAPADIMITRIOU, J. and STOKER, M. G. P. (1971) *Nature New Biol.* **230,** 114-117.
TODARO, G. J. and GREEN, H. (1963) *J. Cell Biol.* **17,** 299-313.

GROWTH CONTROL BETWEEN DISSIMILAR CELLS IN CULTURE

R. A. Weiss and Dorothy L. Njeuma*

Department of Anatomy and Embryology, and Department of Zoology, University College, London

The control of cell proliferation in homogeneous populations is complex enough, and it may appear only to confound the subject further to discuss growth control between different kinds of cell. However, interactions between different cell types occur commonly *in vivo*, and if growth control in cell cultures is relevant to the behaviour of cells in their natural habitat, the growth of cells in mixed cultures merits discussion, particularly the interactions occurring between normal and neoplastic cells. Unnatural mixtures of cells with different properties may also be useful in determining separate steps in growth-controlling mechanisms. For example, in studying density dependent inhibition of growth, mixed cultures enable us to distinguish between acting and responding systems.

In studying cell proliferation in culture we are really examining the population dynamics of a simple community; the Petri dish is our ecosystem. We mention this not as a trite comparison but to emphasize that ecologists' theories of the regulation of numbers of whole organisms are applicable to cell cultures. "Ecology is the study of systems at a level in which individuals or whole organisms may be considered elements of interaction, either among themselves or with a loosely organised environmental matrix" (Margalef 1968). Ecosystems may be treated as cybernetic systems, particularly in the interaction between two or more species or cell types (Slobodkin 1961; Margalef 1968). Negative feedback mechanisms usually restrict population growth before the nutrient supply is so depleted that the organisms starve. This is generally true of cell cultures too, though it is common experience that some neoplastic cells in culture continue to grow in the absence of replenishment of the medium until the whole population dies. Control may be mediated through metabolic

* *Current address:* Faculté des Sciences, Université Fédérale du Cameroun, Yaounde, Cameroun.

inhibitors, such as alcohol in yeast cultures (Gause 1934) or the conditioning factor operating in density dependent proliferation of flour beetles (Park 1955). Population control through epideictic behaviour has been proposed for many animal species (Wynne-Edwards 1962) and analogous theories of census-taking through direct communication between cells in culture are currently popular (e.g. Borek, Higashino and Loewenstein 1969; Bard and Elsdale 1971).

The proliferation of two cell types in mixed culture may theoretically proceed in several ways:

(1) *No interaction*

Each cell type grows independently of the other to reach the same stationary density as it does in "pure" culture. For cell types A and B, where N is the total population,

$$N_{(A+B)} = N_A + N_B$$

We do not know of such a case, but it would be expected in mixed cultures of two cell types whose growth was exclusively controlled by tissue-specific chalones (Bullough 1965), or by specific multiplication-stimulating factors (Temin 1970).

(2) *Growth enhancement*

Two cell types with distinct defects in metabolism may complement each other in mixed culture, for instance through metabolic cooperation, as Pitts (1971) has elegantly shown with BHK cells. Thus

$$N_{(A+B)} > N_A + N_B$$

Non-reciprocal enhancement occurs where only one cell type is metabolically dependent on the other. Growth enhancement between cell types may be quite specific, as in some "inductive" interactions.

(3) *Reciprocal inhibition*

Each cell type "recognizes" the other as part of the total population. Thus,

$$N_{(A+B)} < N_A + N_B$$

Such an interaction may be indirect in the sense that the two cell types do not communicate directly but compete for the same nutrient substances in the culture medium or the same substratum for anchorage. In this case the total population will not exceed that of the cell type with the highest stationary density in "pure" culture,

$$N_A \geqslant N_{(A+B)} \geqslant N_B$$

If space is the limiting factor, the stationary population of the mixed culture will be related to the proportion of the two cell types. This is most easily seen when cells of one type are added to pre-existing clones or islands of the other type.

Where both cell types respond mutually to each other's inhibitory signals, the stationary population of the mixed culture may barely exceed that of the cell type with the lowest stationary density in "pure" culture, as when diploid embryonic fibroblasts are seeded on to stationary monolayers of 3T3 cells (see below, p. 175).

(4) Non-reciprocal inhibition

Njeuma (1971b) has observed non-reciprocal inhibition of growth in mixed cultures of chicken and mouse embryonic fibroblasts. Except in sparse cultures, where there is a mutual feeder action, the mitotic index of chick cells is inversely related to the density of mouse cells, but that of mouse cells appears to be unaffected by the density of chick cells. Inhibition of African green monkey cells by 3T3 cells is also non-reciprocal (Eagle, Levine and Koprowski 1968). Some interactions between normal cells and tumour cells are non-reciprocal.

These models for growth interactions between cells will be discussed in terms of their specificity, and particular attention will be paid to growth-inhibitory mechanisms.

SPECIFICITY IN GROWTH PROMOTION

The specificity of interactions may be studied in mixed cultures or by adding products of one kind of cell to cultures of another. Hormones epitomize highly specific growth regulators between different cells. The specificity resides mainly in the responding tissue, and the use of the term "target cell" is misleading in this respect because it implies a directed stimulus rather than a selective response. Embryonic "inducers", though local in action, are formally similar to hormones and mixed cultures have been usefully exploited in the study of inductive interactions. The growth and differentiation of many types of epithelial cell depend on their cultivation with mesenchymal cells (see Fleischmajer and Billingham 1968). Pancreatic epithelium will proliferate and differentiate in proximity to many kinds of mesenchyme, whereas salivary gland epithelium requires its own mesenchyme. One cell type may provide extracellular materials necessary for the development of another, as in the dependence of myoblasts on collagen (Hauschka and Konigsberg 1966).

The growth-promoting effects of tumours are documented by Argyris

(1966). One specific "growth" stimulating agent is nerve growth factor (Levi-Montalcini 1964). The responding cell types—dorsal root ganglia and sympathetic ganglia—are very restricted, while the emitting tissues are not. Most sarcomata tested, as well as many normal tissues, synthesize or store the factor. It is not clear whether nerve growth factor stimulates the proliferation of neuroblasts. Other, perhaps less specific growth-promoting factors from tumour cells which act on normal cells have been reported (Bürk and Williams 1971; Rubin 1970). If these factors are found to be proteases or lipases we should be careful to distinguish between cell stimulation and cell damage, which frequently induces the proliferation of neighbouring cells. Tumour cells, of course, may destroy normal cells. Mouse sarcoma 180 cells, although enhanced in nerve growth factor, have a toxic effect on fibroblasts (Abercrombie, Heaysman and Karthauser 1957). Chick cells transformed by Rous sarcoma virus produce both toxic and growth-stimulating factors (Rubin 1966, 1970).

At low population densities it is well known that a feeder effect operates between homologous cells (Puck and Marcus 1955; Eagle and Piez 1962; Rothfels, Kupelweiser and Parker 1963; Rubin and Rein 1967; Njeuma 1971a). The feeder effect also operates between cells from different species. Feeder layers of embryonic mouse fibroblasts, for example, have been shown to enhance the growth of BHK21 hamster cells and their polyoma-transformed counterparts (Stoker and Sussman 1965). Sarcoma 180 cell growth was enhanced by embryonic mouse fibroblast feeders, both in suspension and in monolayers (Weiss 1970a). Conditioned medium from dense cultures of mouse or chick fibroblasts stimulated the growth of small numbers of other mouse or chick fibroblasts, irrespective of species, although conditioned medium from mouse fibroblasts was less effective (Rubin 1967). At low population densities chick fibroblasts enhanced the growth of mouse fibroblasts in mixed culture (Njeuma 1971b) and mitosis of chick fibroblasts was stimulated by guinea-pig fibroblasts (Abercrombie, Lamont and Stephenson 1968).

It does not seem, therefore, that feeder effects are species-specific. Whether they are tissue-specific is not clear. Rubin (1967) reported that chick fibroblasts grew better in conditioned medium from chick fibro-blasts than in that from chick kidney epithelial cells.

SPECIFICITY IN GROWTH INHIBITION

Bullough (1965) has proposed that tissue-specific chalones play an important role in the homeostasis of tissues *in vivo*. Growth control by tissue-specific inhibitors was suggested by Bullough and Laurence's (1960) ingenious experiment, in which removal of the epidermis on one side of a mouse's ear led to epidermal proliferation on the other side of the ear as well as at the wound border, but not in the intervening dermis or cartilage. Chalones do not appear to be species-specific. Little work on chalone-like regulators has been done so far using cell cultures, though recent developments in techniques for maintaining differentiated cells *in vitro* (see Ursprung 1968) make this possible.

There is no clear pattern of specificity in density dependent inhibition of cell growth in culture. Eagle and Levine (1967) suggested that growth inhibition may be tissue-specific, fibroblasts inhibiting fibroblasts and epithelial cells inhibiting epithelial cells. While work on such tissue-specificity is scanty, the more numerous observations on species-specificity present a confusing picture. Diploid fibroblasts from various mammalian species, such as mouse, rabbit, monkey and man (Eagle, Levine and Koprowski 1968), are sensitive to density dependent growth, regardless of species. BHK21 cells and their polyoma-transformed variants are inhibited by dense sheets of mouse embryo fibroblasts (Stoker 1964; Stoker, Shearer and O'Neill 1966). 3T3 cells, which are extremely sensitive to homologous inhibition (Todaro and Green 1963), are not inhibited by African green monkey cells (Eagle, Levine and Koprowski 1968). On the other hand, Borek and Sachs (1966) reported that a series of hamster and rat cells transformed by various carcinogens were all subject to heterologous but not homologous inhibition, and they attributed this to a form of allogeneic inhibition. Borek and Sachs' results lack independent corroboration and their carbon-labelling technique for identifying cells in mixed culture is known to be unreliable in the absence of stringent controls (Stoker 1967). Allogeneic inhibition has not been demonstrated in density dependent growth of other fibroblast systems (Pontén and MacIntyre 1968; Pontén and Westermark 1971).

In contrast to mutual interspecific inhibition among diploid mammalian fibroblasts, non-reciprocal inhibition has been found between mammalian and avian cells (Abercrombie, Lamont and Stephenson 1968; Njeuma 1971b). Both mouse and chick embryonic fibroblasts are subject to local density dependent inhibition of growth (Njeuma 1971a), yet in mixed culture mouse fibroblasts inhibited the growth of chick fibroblasts but mouse fibroblasts were not inhibited by chick fibroblasts (Njeuma 1971b). A given density of mouse fibroblasts was not more effective than an equivalent density of chick fibroblasts in inhibiting chick fibroblast mitosis. Therefore in order to explain the non-reciprocity it seems necessary to invoke at least two inhibiting influences, as illustrated in Fig. 1.

The model requires that influence *A* should be different from influence

C. Influence B can be the same as either A or C. If A is similar to B, mouse fibroblasts exert only one influence (A) and chick fibroblasts exert the other (C); but mouse fibroblasts will only respond to A, whereas chick fibroblasts will respond to both A and C. If B is the same as C, mouse fibroblasts must exert two influences, A and C, but will only respond to A, while chick fibroblasts exert C and respond to C from mouse fibroblasts as well as their own. It is of interest to note that mouse and chick fibroblasts in mixed culture do show reciprocal contact inhibition of movement (Abercrombie, Lamont and Stephenson 1968).

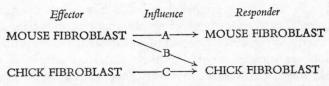

FIG. 1. Model for non-reciprocity of mitotic inhibition in mixed culture.

GROWTH INHIBITION BETWEEN CELLS WITH DIFFERENT TERMINAL DENSITIES

A useful method for assaying the sensitivity of cells to growth inhibition is to add the cells to pre-existing stationary sheets of other cells. If small numbers of cells are added they should not influence the overall culture environment significantly, and the effect of the pre-existing cell sheet on the test cells may be measured. This method was first used by Stoker (1964) to study the effect of diploid fibroblasts on the proliferation of polyoma cells. Where the added cells are morphologically distinct from the pre-existing cell sheet, inhibition of growth can be measured by assaying the number or size of discrete colonies (Stoker 1964; Weiss 1970a). If colony assays are not suitable, the growth of the added cells may be estimated indirectly by subtracting the total cell count of control cell sheets from the total cell count of the mixed cultures (Stoker, Shearer and O'Neill 1966; MacIntyre and Pontén 1967; Eagle, Levine and Koprowski 1968). For several types of neoplastic cell seeded on to normal cell sheets, advantage may be taken of the ability of the neoplastic cell, but not the normal cell, to grow in agar suspension culture, which provides a selective assay of the growth of the neoplastic cells (Stoker, Shearer and O'Neill 1966; Weiss 1970a). Labelling test cells with pigmented particles (Borek and Sachs 1966) or with radio-isotopes may also be useful for identifying cells in mixed culture, provided that great care is taken that the label is maintained in and remains confined to its own cells (Stoker 1967).

In general, mixtures of normal fibroblasts inhibit each other, with the exceptions mentioned in the previous section. For instance, different fibroblastic strains of human cells which were susceptible to density inhibition of growth in pure culture, usually inhibited each other in mixed culture (Eagle and Levine 1967; Bard and Elsdale 1971). In this type of mutual inhibition, cells with a low terminal density tend to exert a dominant influence over cells with higher terminal densities, provided that the latter cells are seeded sparsely. Thus, the growth of embryonic mouse or chick fibroblasts is completely inhibited by seeding them sparsely on to stationary 3T3 cells, although the same density of embryonic fibroblasts proliferates actively in pure culture, even in medium taken from stationary 3T3 cultures (Weiss, unpublished observations). The embryonic fibroblasts appear to be arrested in the G1 phase of the cell cycle and 3T3 cells may be used to synchronize the embryonic fibroblasts in that phase. Similarly, Pontén and Westermark (1971) have used stationary cultures of human glial cells, which have a terminal density comparable to that of 3T3 cells, to inhibit the proliferation of human fibroblasts (though lung fibroblasts were not completely inhibited). Several malignant glioma cell lines proliferated to varying extents on the stationary glial monolayers. Apart from its use in discerning interactions between normal cells and tumour cells, this method is proving useful in suppressing the proliferation of stromal cells in primary tumour cultures (Aaronson, Todaro and Freeman 1970).

Stationary cell sheets have been used mostly to study the influence of normal cells on the proliferation of tumour cells. Stoker (1964) showed that BHK21 cells transformed by polyoma virus were susceptible to growth inhibition by dense sheets of mouse embryo cells, although the polyoma cells grew to a very high density in pure culture. The sensitivity of polyoma-transformed cells to density inhibition by normal cells was confirmed by Pontén and MacIntyre (1968) using bovine polyoma-transformed and normal cells and by Weiss (1970a) using Stoker's system. However, polyoma-transformed variants have been found which are not sensitive to heterologous inhibition by normal cells or by 3T3 cells (Pollack, Green and Todaro 1968; Weiss, see below, p. 176). Cells transformed by Rous sarcoma virus lose their density dependent growth, whether they are derived from chicken, hamster, rat or bovine cells (Weiss 1970a; MacIntyre and Pontén 1967). Similarly, SV40-transformed 3T3 cells, providing their transformation has been fully expressed, are no longer susceptible to inhibition by untransformed 3T3 cells (Todaro and Green 1966). Recent experiments showed that mouse sarcoma 180 cells, which appeared to be completely suppressed by dense mouse embryo cell sheets (Weiss 1970a), in fact grew slowly (unpublished observations), as they did on African green monkey cells (Eagle, Levine and Koprowski 1968). Retardation of growth rather than complete inhibition may be common among neoplastic cells in contact with normal cell sheets (Pontén and Westermark 1971; Weiss and Veselý, unpublished observations).

The inhibition of growth of polyoma cells on heterologous cell sheets appears to depend both on density and on the non-proliferative state of the underlying cell sheet. Sparsely seeded normal cells which are non-proliferating, such as X-irradiated cells, make excellent feeder layers rather than inhibitory layers; and dense cell sheets which are still proliferating, such as untransformed BHK21 cells, are also poor inhibitors. A surprising finding was that X-irradiated polyoma cells inhibited the growth of non-irradiated polyoma cells (Stoker 1964). It would be interesting to know whether X-irradiation induced the formation of giant cells, as giant polyoma cells are known to be inhibitory, at least to cell locomotion (Stoker 1964).

Our own studies on the inhibition of polyoma cells suggested that some proliferation took place among dense mouse fibroblast sheets. Colony-inhibition tests showed that this incomplete inhibition was due to a small proportion of polyoma cells growing into large colonies (Weiss 1970a). It was of interest to determine whether this was due to selection of variants which were unresponsive to heterologous growth inhibition or to the ecological situation of those cells in the Petri dish. This was tested by replating the colonies which developed on to new sheets of dense fibroblasts. Ten out of eleven colonies tested in this way yielded a similar, low proportion of colonies when seeded on to dense cell sheets, indicating that there was no permanent change in the responsiveness of the polyoma cells. The eleventh colony behaved as a permanently altered clone no longer sensitive to heterologous growth inhibition. We noticed that colony inhibition appeared to be less marked when the polyoma cells were inefficiently dispersed during trypsinization. Therefore, polyoma cells were allowed to clump in a shaker flask and then seeded on to inhibitory cell sheets. There was a marked diminution in colony inhibition, but it was difficult to quantify because of the fragility and variable size of the aggregates, and because of the presence of single polyoma cells which formed colonies on the control cell sheets which were seeded with sparse polyoma cells. An alternative approach was to seed monodispersed polyoma cells into empty dishes and to add excess normal cells after different time intervals. Relative colony inhibition decreased with increasing time in culture before the inhibitory fibroblasts were seeded. It appeared that the critical clone size to overcome growth inhibition was between 8 and 32 cells. This estimate is imprecise because of variability in the rate of growth of polyoma clones and in the spatial dispersion of cells within one clone. It is possible that smaller clones overcome growth inhibition, provided that the polyoma cells maintain their homologous contacts. The mouse fibroblasts may act by transporting normal cell surface components to the polyoma cells, by analogy with the density dependence imposed by a coat of concanavaline A (Burger 1971).

MECHANISMS OF DENSITY DEPENDENT GROWTH INHIBITION

What conclusions can we draw about the behaviour of malignant cells and the inhibition of their growth by normal cells? Is sensitivity to heterologous inhibition associated with a low terminal density in pure culture? Such a correlation was suggested by Eagle, Levine and Koprowski (1968) and by Pollack, Green and Todaro (1968), but was not found for human glioma cells (Pontén and Westermark 1971) or for the BHK21 polyoma cells discussed above. This disparity between homologous and heterologous growth inhibition led Stoker (1964) to suggest that whereas normal cells both emit and receive inhibitory signals, polyoma cells no longer emit them but can still respond to them. Under this scheme, Rous cells appear to have lost the ability to respond to signals from normal cells. Their emitting function has not been studied in detail but observations by Weiss and Veselý (unpublished) and MacIntyre (personal communication) indicate that dense sheets of Rous-transformed mammalian cells retard the growth of untransformed cells and polyoma-transformed cells. Thus, there is non-reciprocal inhibition in mixed cultures of normal cells and mammalian Rous-transformed cells. This might be for a "trivial" reason. The growth of normal fibroblasts is anchorage dependent (Stoker et al. 1968), while that of Rous cells is not (Weiss 1970a); yet Rous cells, given the opportunity, will occupy space on the substratum for which normal cells compete unsuccessfully, because Rous cells contact-inhibit the movement of normal cells (Veselý and Weiss 1971).

One might expect heterologous density dependent growth to be associated with anchorage dependence, on the hypothesis that inhibitions of cell growth in suspension and in dense layers are both due to the inhibition of cell spreading (Castor 1968). But BHK21 polyoma cells grow well in suspension (Stoker et al. 1968), so their sensitivity to growth inhibition by anchored normal cells suggests that some positive inhibitory influence is exerted (e.g. donation of cell surface components; see above). The distinction between serum dependence and "topoinhibition" is relevant here (Dulbecco 1971), since anchorage dependence and serum dependence appear to be closely related. Polyoma cells grown in suspension at high densities ($> 2 \times 10^6$ cells/ml) appear to produce a growth inhibitor (Bellanger, Jullien and Harel 1970). However, growth is not inhibited in suspension at densities equivalent to those in layer cultures, nor between layered cells and suspended cells (Weiss 1970a), nor across a thin Millipore filter (Schutz and Mora 1968). Density dependent growth may, therefore, be mediated through cell contact, and the low-resistance junction (Furshpan and Potter 1968; Borek, Higashino and Loewenstein 1969) is a popular candidate for the transmission of growth control. Since the low resistance junction, unlike the nerve synapse, is not a rectifying junction, the presence of junctional coupling does not explain non-reciprocal growth inhibition. Low-resistance junctions may be a pre-requisite for density dependent growth inhibition, but that does not mean that all coupled cells necessarily respond to the growth inhibitory influence.

The susceptibility of tumour cells to growth inhibition by normal cells *in vitro* would be expected to be correlated with tumorigenicity *in vivo*, as the experiments of Pollack, Green and Todaro (1968) suggest. However very few systems have been developed where the behaviour of cells *in vitro* is being systematically investigated together with *in vivo* studies in syngeneic hosts (Aaronson and Todaro 1968; Weiss and Veselý, unpublished), and more serious thought must be given to *in vivo* assays and ways of quantifying them.

ROLE OF MITOSIS IN THE PROCESS OF CELL TRANSFORMATION

So far our discussion has been concerned with growth interactions between overtly different cell types. We wish now to comment on the dependence of cell differentiation and transformation on mitosis. There is increasing evidence that the transformation of cells by many tumour viruses depends on the cell division cycle. The transformation of chick fibroblasts by Rous sarcoma virus (RSV) is inhibited by X-rays (Rubin and Temin 1959), by inhibitors of DNA synthesis, such as cytosine arabinoside (Bader 1966), or by blocking the cell division cycle with low serum and excess thymidine (Temin 1967). Weiss (1970b) studied the effect of crowded monolayers of normal cells on the transformation of chick cells freshly infected with RSV. It was found that dense sheets of mouse embryo cells inhibited the transformation of infected chick cells, though dense chick or quail cells were not themselves inhibitory, whether or not they were susceptible to the strain of RSV used to infect the challenge cells (Weiss 1971). Similarly, non-proliferating sheets of mouse cells are resistant to transformation by murine sarcoma virus

(Yoshikura 1970; Bather and Leonard 1970) and SV40 (Todaro and Green 1966). Thus, infection with these tumour viruses does not *induce* cell replication in crowded cultures although transformation depends on cell division. On the other hand, infection with RSV of cells in agar suspension induces transformation (Weiss 1970b) and infection with polyoma virus of mouse cells which are stationary in medium containing a low serum concentration induces DNA synthesis (Dulbecco, Hartwell and Vogt 1965; Fried and Pitts 1968). In the latter case it is not clear whether the lack of inhibition is related to the lytic outcome of infection or to the method by which the cells are maintained in a non-proliferating state.

Since Rous-transformed cells are not subject to density dependent growth inhibition while freshly infected cells are, the time after infection at which the cells cease to respond to growth inhibition could be studied (Weiss 1970b), and it was found that nearly all infected cells became independent of density within twelve hours of infection. An experiment on the loss of density dependent growth inhibition following RSV infection is shown in Fig. 2. In this case, chick fibroblasts were synchronized and inhibited in the G1 phase of the cell cycle by seeding them on to static 3T3 monolayers for 48 hours. They were then resuspended, infected with RSV and reseeded at a subconfluent density to promote the cell division cycle. At different time intervals after infection sample cultures were trypsinized once more and equal aliquots of cells were seeded on to confluent and sparse 3T3 monolayers and overlaid with agar medium four hours later. If the infected cells were reseeded on to static 3T3 cultures within six hours of infection, transformation was completely suppressed, but if contact with static 3T3 cells was delayed for eighteen hours, transformation was not significantly inhibited. Nakata and Bader (1968), using X-irradiation to inhibit growth, similarly found that the morphological conversion of cells infected with RSV became "fixed" within eighteen hours of infection. Using SV40-infected 3T3 cells, Todaro and Green (1966) found that one cell division was enough to "fix" the virus in the cell, but that further divisions were required before transformation was expressed as density independent growth.

The dependence of transformation by tumour viruses on the cell cycle appears in most cases to be independent of the stabilization of the viral genome (Todaro and Green 1966; Weiss 1970b) and the formation of provirus (Temin 1967). However, viral replication is suppressed concomitantly with cell transformation. Temin (1967) and Hobom–Schnegg, Robinson and Robinson (1970) showed by using synchronized cultures that maturation of RSV in chick cells does not occur until after the next mitosis following infection.

The dependence of viral transformation and replication on cell division has striking parallels in normal cell differentiation. For example, the earliest response of pancreatic epithelium to "induction" by mesenchyme is DNA synthesis and mitosis (Wessels 1964), and hormone-induced

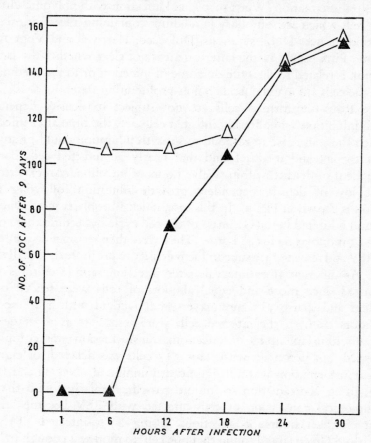

FIG. 2. Loss of density dependent growth inhibition of chick fibro-blasts following infection with Rous sarcoma virus. The infected cells were seeded on to confluent sheets (full triangles) or sparse sheets (open triangles) of 3T3 cells at the times indicated and foci of Rous cells were counted 9 days later.

mammary gland differentiation is also preceded by mitotic stimulation (Turkington 1968). Holtzer and co-workers found that differentiation of myoblasts and chondrocytes *in vitro* was preceded by a critical cell division cycle (Bischoff and Holtzer 1969; Holtzer and Abbott 1968). Indeed, Nameroff and Holtzer (1969) showed that myogenesis may be inhibited by seeding chick embryo myoblasts on to confluent, non-proliferating layers of homotypic or heterotypic cells. Like cell transformation by RSV (Weiss 1970*b*), myogenesis took place after diluted replating of the mixed culture, and like transformation of cells by murine sarcoma virus (Yoshikura 1970), myotubes developed in a wound made in the crowded sheet. Thus there is a strong argument for interpreting *in vitro* neoplastic transformation of cells and replication of virus as essentially similar to cell differentiation, in which the synthesis of new proteins depends on a critical or "quantal" mitosis (Bischoff and Holtzer 1969).

Low concentrations of 5-bromodeoxyuridine (BUdR) reversibly inhibit cell differentiation in chondrogenesis and myogenesis, without affecting cell proliferation (Holtzer and Abbott 1968; Lasher and Cahn 1969; Coleman, Coleman and Hartline 1969; Bischoff and Holtzer 1970). These findings led Holtzer and co-workers to propose that there are two categories of gene expression during differentiation. First is the activity of "household" genes, common to all cells, coding for basic metabolic machinery such as respiratory enzymes and molecules required for cell proliferation. Second is the activity of tissue-specific "luxury" genes, thought to control differentiation. BUdR at low concentrations appears to affect only this second category. In such a scheme, should we consider neoplastic transformation and viral replication as "household" or "luxury" events? Silagi and Bruce (1970) recently reported that both malignancy and differentiation were suppressed in melanotic melanoma cells treated with BUdR. The necessity of "quantal" mitosis for the expression of RNA tumour viruses suggests that cell transformation and viral replication should be reversibly suppressed by BUdR, but there is little evidence to support this view.

SUMMARY

Some models are presented of the possible interactions among different cell types cultivated within the same ecosystem. These include competition between cells for space or nutrients, and modes of communication between different cells which may play a part in population control. The specificity of interactions between dissimilar cells in culture is discussed. Nonreciprocal inhibition of growth is revealed in mixed cultures of chick and mouse fibroblasts where the proliferation of chick cells is dependent on the density of mouse cells but that of mouse cells is independent of chick cells.

Most tumour cells, with the notable exception of polyoma-transformed cells, proliferate on inhibited sheets of normal cells, although the rate of proliferation may be severely retarded by the normal cells. Polyoma cells, and normal cells with a high terminal density in pure culture, are inhibited in their growth by cells with lower terminal densities in conditions where there is little or no homologous contact between the high-density cells.

Cells transformed by Rous sarcoma virus continue to proliferate when seeded among crowded, non-proliferating mouse fibroblasts but the transformation of freshly infected cells is suppressed in these conditions. When the mitotic cycle of the infected cells is not immediately inhibited it can be shown that they lose their capacity to respond to the growth inhibitory influence of normal fibroblasts within twelve hours of infection, revealing an earlier change in growth control than has hitherto been described. A theory current among developmental biologists, that a cell must proceed through a critical division cycle before a new pattern of protein synthesis occurs, may be applied to the transformation of cells by Rous sarcoma virus and other tumour viruses.

Acknowledgement

We are grateful to Michael Abercrombie for his interest and advice.

REFERENCES

AARONSON, S. A. and TODARO, G. J. (1968) *Science* **162**, 1024–1026.

AARONSON, S. A., TODARO, G. J. and FREEMAN, A. E. (1970) *Expl Cell Res.* **61**, 1–5.

ABERCROMBIE, M., HEAYSMAN, J. E. M. and KARTHAUSER, H. M. (1957) *Expl Cell. Res.* **13**, 276–291.

ABERCROMBIE, M., LAMONT, D. M. and STEPHENSON, E. M. (1968) *Proc. R. Soc. B* **170**, 349–360.

ARGYRIS, T. (1966) *Adv. Biol. Skin* **7**, 55–73.

BADER, J. P. (1966) *Virology* **29**, 444–451.

BARD, J. and ELSDALE, T. (1971) This volume, pp. 187–197.

BATHER, R. and LEONARD, A. (1970) *J. gen. Virol.* **7**, 249–256.

BELLANGER, F., JULLIEN, M. and HAREL, L. (1970) *C. r. hebd. Séanc. Acad. Sci., Paris,* *Sér. D* **270**, 2232–2235.

BISCHOFF, R. and HOLTZER, H. (1969) *J. Cell Biol.* **41**, 188–200.

BISCHOFF, R. and HOLTZER, H. (1970) *J. Cell Biol.* **44**, 134–150.

BOREK, C., HIGASHINO, S. and LOEWENSTEIN, W. R. (1969) *J. Membrane Biol.* **1**, 274–293.

BOREK, C. and SACHS, L. (1966) *Proc. natn. Acad. Sci. U.S.A.* **56**, 1705–1711.

BULLOUGH, W. S. (1965) *Cancer Res.* **25**, 1683–1727.

BULLOUGH, W. S. and LAURENCE, E. B. (1960) *Proc. R. Soc. B* **151**, 517–536.

BURGER, M. M. (1971) This volume pp. 45–63.

BÜRK, R. R. and WILLIAMS, C. A. (1971) This volume, pp. 107–120.

CASTOR, L. N. (1968) *J. cell. Physiol.* **71**, 161–172.

COLEMAN, J. R., COLEMAN, A. W. and HARTLINE, J. H. (1969) *Devl Biol.* **19**, 527–548.

DULBECCO, R. (1971) This volume, pp. 71–76.

DULBECCO, R., HARTWELL, L. H. and VOGT, M. (1965) *Proc. natn. Acad. Sci. U.S.A.* **53**, 404–410.

EAGLE, H. and LEVINE, E. M. (1967) *Nature, Lond.* **213**, 1102–1106.

EAGLE, H., LEVINE, E. M. and KOPROWSKI, H. (1968) *Nature, Lond.* **220**, 266–269.

EAGLE, H. and PIEZ, K. (1962) *J. exp. Med.* **116**, 29–43.

FLEISCHMAJER, R. and BILLINGHAM, R. (1968) *Epithelial Mesenchymal Interactions.* Baltimore: Williams & Wilkins.

FRIED, M. and PITTS, J. D. (1968) *Virology* **34**, 761–770.

FURSHPAN, E. J. and POTTER, D. D. (1968) *Curr. Top. Dev. Biol.* **3**, 95–127.

GAUSE, G. F. (1934) *The Struggle for Existence.* Baltimore: Williams & Wilkins.

HAUSCHKA, S. and KONIGSBERG, I. R. (1966) *Proc. natn. Acad. Sci. U.S.A.* **55**, 119–126.

HOBOM-SCHNEGG, B., ROBINSON, H. L. and ROBINSON, W. S. (1970) *J. gen. Virol.* **7**, 85–93.

HOLTZER, H. and ABBOTT, J. (1968) In *Stability of the Differentiated State*, pp. 1–16, ed. Ursprung, H. Heidelberg: Springer.

LASHER, R. and CAHN, R. D. (1969) *Devl Biol.* **19**, 415–435.

LEVI-MONTALCINI, R. (1964) *Science* **143**, 105–110.

MACINTYRE, E. and PONTÉN, J. (1967) *J. Cell Sci.* **2**, 309–322.

MARGALEF, R. (1968) *Perspectives in Ecological Theory.* Chicago: University of Chicago Press.

NAKATA, Y. and BADER, J. P. (1968) *Virology* **36**, 401–410.

NAMEROFF, M. and HOLTZER, H. (1969) *Devl Biol.* **19**, 380–396.

NJEUMA, D. L. (1971a) *Expl Cell Res.* in press.

NJEUMA, D. L. (1971b) *Expl Cell Res.* in press.

PARK, T. (1955) In *The Numbers of Man and Animals*, ed. Cragg, J. B. and Pirie, N. W. London: Oliver and Boyd.

PITTS, J. D. (1971) This volume, pp. 89–98.

POLLACK, R., GREEN, H. and TODARO, G. J. (1968) *Proc. natn. Acad. Sci. U.S.A.* **60**, 126–133.

PONTÉN, J. and MACINTYRE, E. H. (1968) *J. Cell Sci.* **3**, 603–613.

PONTÉN, J. and WESTERMARK, B. (1971) *Abstracts X Int. Cancer Res. Congress*, Huston.

PUCK, T. T. and MARCUS, P. I. (1955) *Proc. natn. Acad. Sci. U.S.A.* **41**, 432–437.

ROTHFELS, K. H., KUPELWEISER, E. B. and PARKER, R. C. (1963) *Proc. Can. Cancer Conf.* **5**, 191–223.

RUBIN, H. (1966) *Expl Cell Res.* **41**, 149–161.

RUBIN, H. (1967) In *The Specificity of Cell Surfaces*, pp. 181–194, ed. Davis, B. D. and Warren, L. Englewood Cliffs, N.J.: Prentice Hall.

RUBIN, H. (1970) *Science* **167**, 1271–1272.

RUBIN, H. and REIN, A. (1967) In *Growth Regulating Substances for Animal Cells in Culture*, pp. 51–66, ed. Defendi, V. and Stoker, M. Philadelphia: The Wistar Institute Press.

RUBIN, H. and TEMIN, H. M. (1959) *Virology* **7**, 75–91.

SCHUTZ, L. and MORA, P. T. (1968) *J. cell. Physiol.* **71**, 1–6.

SILAGI, S. and BRUCE, S. A. (1970). *Proc. natn. Acad. Sci. U.S.A.* **66**, 72–78.

SLOBODKIN, L. B. (1961) *Growth and Regulation of Animal Populations.* New York: Holt, Rinehart & Winston.

STOKER, M. G. P. (1964) *Virology* **24**, 165–174.

STOKER, M. G. P. (1967) *J. Cell Sci.* **2**, 293–304.

STOKER, M. G. P., O'NEILL, C., BERRYMAN, S. and WAXMAN, V. (1968) *Int. J. Cancer* **3**, 683–693.

STOKER, M. G. P., SHEARER, M. and O'NEILL, C. (1966) *J. Cell Sci.* **1**, 297–310.

STOKER, M. G. P. and SUSSMAN, M. (1965) *Expl Cell Res.* **38**, 645–653.

TEMIN, H. M. (1967) *J. cell. Physiol.* **69**, 53–64.

TEMIN, H. M. (1970) In *Biology of Large RNA Viruses*, ed. Barry, R. and Mahy, B. London: Academic Press.

TODARO, G. J. and GREEN, H. (1963) *J. Cell Biol.* **17**, 299–313.

TODARO, G. J. and GREEN, H. (1966) *Proc. natn. Acad. Sci. U.S.A.* **55**, 302–308.

TURKINGTON, R. W. (1968) *Curr. Top. Dev. Biol.* **3**, 199–218.

URSPRUNG, H. (ed.) (1968) *Stability of the Differentiated State.* Heidelberg: Springer.

VESELÝ, P. and WEISS, R. (1971) *Microcinematography as a Research Method in Cytology*, ed. Hroch, M. Prague: Charles University. In press.

WEISS, R. A. (1970a) *Expl Cell Res.* **63**, 1–18.

WEISS, R. A. (1970b) *Int. J. Cancer* **6**, 333–345.

WEISS, R. A. (1971) *Virology* in press.
WESSELS, N. K. (1964) *J. Cell Biol.* **20**, 415–433.
WYNNE-EDWARDS, V. C. (1962) *Animal Dispersion in Relation to Social Behaviour.* Edinburgh: Oliver & Boyd.
YOSHIKURA, H. (1970) *J. gen. Virol.* **6**, 183–185.

DISCUSSION

Stoker: Perhaps we should ask Harry Rubin to explain non-reciprocal inhibition on the basis of pH!

Rubin: Non-reciprocal inhibition is just as easy or as hard to explain in the pH model as it is in any other. One need only make the reasonable assumption that the mouse and chicken cells have different degrees of sensitivity to pH to get a different response when they are interacting with each other.

I should like to ask Dr Weiss how quantitative his data are: he gave no figures in his talk. I am hesitant about using the notion of growth inhibition in a qualitative sense with chicken cells. I find nothing approaching an absolute cessation of growth unless the medium is depleted. Even when one seeds trypsinized cells so that the equivalent of more than two monolayers attach, many of the cells will undergo division within a fairly short time. This would complicate the interpretation of experiments in which trypsinized cells are plated in crowded cultures and observations made for only a limited period. Could you describe the methodology in more detail?

Weiss: This is Dr Njeuma's work (1971). Her standard method was to seed the cells together, fix the cultures 48 hours after seeding and 3 hours after treatment with colcemid, and calculate the mitotic rate. But she also studied the mitotic rate at different times after setting up the cultures and at various initial densities.

Stoker: Can you distinguish the two cell types in mitosis?

Weiss: This is the trickiest part, but it can be done reliably.

Rubin: It's no problem to handle the non-reciprocal effect, by setting up different thresholds for sensitivity to whatever alters when the cells are put together. You can have perfectly simple models.

Weiss: A given density of mouse cells has the same inhibitory effect on chick cells as the equivalent density of chick cells but that same density of chick cells does not inhibit the mouse cells.

Rubin: No matter what the model is, one has to say that chicken and mouse cells interact differently in heterologous and in homologous combinations. By setting different thresholds of sensitivity to local pH of the two cell types, and saying that chick–mouse does not approach as closely as chick–chick, I could give you as good or bad an explanation under the pH model as under any other.

Shodell: Isn't the change in ψ potential dependent only upon the distance between cells? That is, shouldn't the change in ψ be a strict function solely of cell density?

Rubin: There is indeed an interaction, and this determines how closely, in atomic dimensions, the cell surfaces can approach one another. But density is measured at a relatively crude level. When I speak of interactions which affect surface potential and local pH I refer to separations of 10 Å or so, and this requires specificity. By contrast, consider the distance between the cell and its glass or plastic substratum. Superficially the cell appears closely adherent to the dish, but actually most of its surface is a good distance away from the dish.

Bard: If your theory can explain anything, Dr Rubin, what would you consider a helpful experiment in establishing its truth?

Rubin: Of course, any theory which can explain everything is a bad theory. It's just that non-reciprocal inhibition is not a proper test for the pH model of control. There are many experiments which are, including the use of fluorescent probes to determine interfacial pH and the use of cationic and anionic polyelectrolytes to alter it, and to observe effects on growth.

Abercrombie: Dr Njeuma's work was mainly concerned with the relation between the contact inhibition of movement and the contact inhibition of mitosis, or topoinhibition. Chick cells are very effective in contact-inhibiting the movement of mouse cells but they are quite ineffective at inhibiting their mitosis, and these experiments were intended to show that these two phenomena are distinct. Her work makes this point.

Stoker: Cross-inhibition between different cell types only seem to work in special cases; for example, you require mouse (or rat) cells to inhibit other cells, and I think the experiments only worked out if the inhibited cell was a polyoma-transformed, or an early Rous-transformed cell.

Weiss: Among transformed, or neoplastic cells only polyoma cells are completely inhibited. I thought that S180 mouse sarcoma cells were too, but although their growth is enormously retarded, it goes on. Many kinds of non-transformed cells with high homologous saturation densities are inhibited by cells with lower saturation densities.

Dulbecco: Is there a relationship between the pattern of heterologous interaction of cells and the pattern of the heterologous effect of serum on cells?

Weiss: I don't think so. Are mammalian sera generally bad for chick cells and good for other mammalian species, Dr Rubin?

Rubin: Observations on species-specificity of sera for cell growth have a long history in tissue culture (Fischer 1946). More recently Holley and Kiernan (1968) and I (this volume, pp. 127–145) have reported that cells do best in their homologous sera.

Burger: Dr Weiss mainly talked about the interaction between normal and transformed cells; what about the interaction between transformed cells and other transformed cells? Borek and Sachs (1966) studied the interaction between virally, chemically and X-ray transformed cells and showed clearly that such pairs in general were capable of inhibiting each other.

Weiss: Borek and Sachs (1966) claimed that the growth of transformed cells was inhibited by any other cell type, including other types of transformed cells. As we mentioned in the text (p. 173), we doubt the reliability of their results. They supposed that because different transformed cells have specific antigens on the cell surface they might be inhibited by cells which lack these antigens. They suggested that this might be a form of allogeneic inhibition (Hellström and Hellström 1965; Möller and Möller 1965). In fact, this is the reverse argument to allogeneic inhibition, in which cells possessing certain antigens are cytotoxic to cells which lack them. Concerning interactions between transformed cells, E. MacIntyre (unpublished data) studied the effect of Rous-transformed bovine cells on polyoma-transformed bovine cells (both syngeneic and allogeneic) and she found partial inhibition of the polyoma cells. P. Veselý and I (unpublished data) studied the interactions of spontaneously transformed LW13 rat cells and LW13 cells additionally transformed by Rous sarcoma virus. Crowded layers of one cell type slowed the cell cycle time of the other, but proliferation was not completely inhibited.

REFERENCES

BOREK, C. and SACHS, L. (1966) *Proc. natn. Acad. Sci. U.S.A.* **56**, 1705–1711.
FISCHER, A. (1946) *The Biology of Tissue Cells.* Copenhagen: Glydendalske Boghandel Nordisk Forlag.
HELLSTRÖM, K. E. and HELLSTRÖM, I. (1965) *Wistar Inst. Symp. Monogr.* **3**, 79–91.
HOLLEY, R. W. and KIERNAN, J. A. (1968) *Proc. natn. Acad. Sci. U.S.A.* **60**, 300–304.
MÖLLER, G. and MÖLLER, E. (1965) *Nature, Lond.* **208**, 260.
NJEUMA, D. L. (1971) *Expl Cell Res.* in press.

SPECIFIC GROWTH REGULATION IN EARLY SUBCULTURES OF HUMAN DIPLOID FIBROBLASTS

JONATHAN BARD AND TOM ELSDALE

Medical Research Council Clinical and Population Cytogenetics Unit, Western General Hospital, Edinburgh

A FRESHLY initiated culture of fibroblasts grows at an exponential rate until there are enough cells to form a continuous sheet (confluence), at which point growth may cease or continue at a reduced rate. It is the mechanism by which cells *in vitro* recognize differences in cellular environment and so alter their division cycle which is known as growth regulation. The phenomenon may be considered from two distinct viewpoints. The first examines it in the light of viral transformation of cells—the growth pattern may alter markedly after transformation—and concentrates attention on the genotype. The second concerns itself more with the phenotypic aspects, as it is hoped that insight into the phenomenon *in vitro* will be helpful in understanding growth processes *in vivo*. We have been concerned with this latter approach.

There are three types of growth pattern observed in cell cultures. Some lines of cells cease growth at confluence. Other lines grow indefinitely in the same culture. The third situation is one intermediate between these extremes, and is illustrated by the behaviour of early subcultures of human, diploid, foetal lung fibroblasts. Techniques for establishing cell strains are described in previous publications (Elsdale 1968; Elsdale and Foley 1969). Under a standard regime of thrice-weekly changes of a medium containing 10 per cent serum, cultures grow to stationary densities of around 10×10^6 cells per 50 mm dish—about six confluence equivalents. This raises the question of how growth is regulated to give final populations of a characteristic degree of overconfluence. Our experiments show that the stationary density is a variable dependent on the conditions of culture. We have counted 35×10^6 cells in a 50 mm plate continuously perfused with a 10 per cent serum medium. However, under standard conditions, cellular specificity defines the growth pattern; adult skin fibroblasts, for example, give stationary densities of around 4×10^5, about half that of lung fibroblasts.

In seeking to understand the growth behaviour of fibroblasts under standard conditions we have asked three questions. Does the distribution and arrangement of the cells within a culture have repercussions for growth regulation? Are cell-specific, growth-inhibiting influences passed through the medium or by cell-to-cell contacts? What is the nature of the substratum supporting the cells?

THE DISTRIBUTION AND ARRANGEMENT OF THE CELLS

Normal fibroblasts have certain pattern-forming propensities that can be realized in culture in different ways (Elsdale 1968; Elsdale and Foley 1969). Post-confluent lung fibroblasts in plastic dishes normally make thick ridge-like structures separated from one another by thinner areas (Fig. 1a). The distribution of cells in a normal stationary culture is therefore markedly uneven. Where, however, they are initiated on a thin and transient aligned collagen substratum, the cells stretch out parallel to one another, and the parallel array persists thereafter, including all the cells produced by subsequent growth. This distribution is quite uniform; there is no significant variation from one region to another within the culture (Fig. 1b). The question arises whether growth is the same or different in cultures of the same strain when the distribution of the cells is uniform or non-uniform. We have failed to show a significant difference in the stationary densities in preliminary experiments. It appears then that we may need to think of growth as dependent on the density of a culture overall—the integral density—and not on local densities. Integral-density-dependent growth regulation implies some mechanism for the averaging of cell states over the culture as a whole.

CELLULAR INTERACTION AND THE AVERAGING OF CELL STATES

There are two possible modes of interaction leading to the averaging of cell states in growth regulation. The first assumes a soluble growth inhibitor, manufactured by the cells, not necessarily by all of them, and mobile within the medium. There is evidence against such an inhibitor from experiments in which wounds in stationary cell sheets are observed to prompt local cell division as part of the mending process. The second assumes that contact interactions are of over-riding importance. We have performed experiments of the following kind that also favour the second possibility.

Cultures are initiated with adult skin fibroblasts that grow to stationary densities of around 4×10^6 cells per 50 mm dish. These are the low density

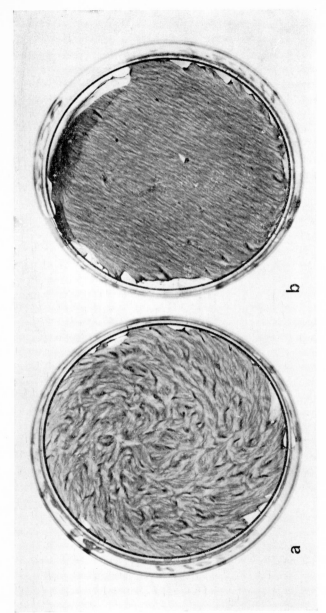

Fig. 1. Embryonic-lung diploid fibroblasts grown to stationary densities in 50 mm Nunclon dishes on (a) plastic and (b) aligned collagen substrates.

[To face page 188

or LD cells, and are allowed to grow to near stationary densities, at which time a circular area 25 mm in diameter in the centre of the dish is freed of cells by scraping with an orange stick. About 4×10^5 lung fibroblasts, of stationary density around 10×10^6 cells per 50 mm dish (high density or HD cells), are added over the whole culture, which is maintained under standard conditions, the border being scraped daily. There are two populations of HD cells in these cultures. Some of the HD cells fall on stationary LD cells and are incorporated into the existing cell sheet; others fall on the central empty region and these proliferate a pure

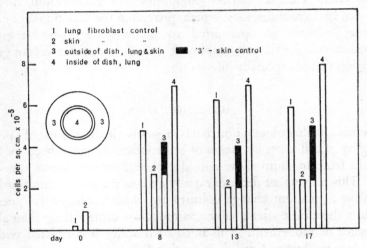

FIG. 2. Counts from a mixed cell experiment, with adult skin (low density) and embryonic lung (high density) fibroblasts on the outer part of the dish, and high density lung cells alone on the central area. The shaded region in (3) shows the excess of growth of the skin-lung mixture over the skin control. (For detailed explanation, see text.)

population of HD cells. By counting whole dishes, and dishes after scraping away the centres, the growth over the two areas in the cultures can be assessed. The results (Fig. 2) show that there is little growth in the mixed populations, whereas the pure populations of HD cells grow to high density, even although they are growing in a medium supporting a larger population of stationary LD cells.

It is clear from this experiment that HD cells are not inhibited from growing by materials released into the medium by stationary LD cells, whereas contact with stationary LD cells is strongly inhibiting, providing the LD cells outnumber the HD cells several fold. The rule emerges that LD cells may inhibit HD cells, but a stimulating effect of HD cells on LD cells has not been demonstrated.

7*

These results clearly disclose the existence of growth-inhibiting influences in fibroblast cultures and show that those mediated via cell contact are more important than medium-borne influences for cell-specific growth regulation under standard conditions. It appears therefore that the averaging of cell states postulated above as a likely condition of growth regulation occurs through cell contacts and not through the medium.

This conclusion does not of course imply that the provision of nutrients cannot be limiting, nor that the alterations in the medium resulting from cell metabolism do not affect cell growth. It is easy to demonstrate effects of this kind. Variations in the composition of the medium, however, serve as a background against which, providing the variations are not too extreme, cultures are permitted to demonstrate growth-regulating patterns specific for the cells employed. It is the specific aspects of growth regulation that are specially interesting.

CELL MOTILITY

There is a further observation that is suggestive in the context of a possible averaging of cell states by means of cell contact. Dense cultures of fibroblasts in fresh medium show considerable cell movement in time-lapse films. This movement declines rapidly and on the second and third days following a medium change cultures are almost completely quiescent. Within a short time after a subsequent medium change there is an abrupt resurgence of movement. This burst of motility is correlated with the often observed burst of cell division that follows a medium change and involves a minority of the cells. Fresh serum added to old medium causes a very little stimulation of cell growth (which according to Griffiths (1970) can be accounted for by the presence of small molecular nutrients in the serum). We observe almost no resurgence of cell movement when fresh serum is added to a quiescent culture in old medium.

If the inhibition of growth in dense cultures is dependent on messages passing between cells via contacts allowing ionic continuity, the result of disturbing the contacts between the cells should be seen as a pulse of growth. Clearly the abrupt resurgence of motility in a previously quiescent culture could create just such a disturbance. We observe correlations between motility and growth, quiescence and growth inhibition. These correlations also provide an explanation of the unlimited growth observed when cells are continuously perfused with fresh medium: sustained motility will not permit stable junctions to form. It is interesting to note that Sefton and Rubin (1970) have induced growth in stationary fibroblast cultures by adding low concentrations of trypsin.

MECHANISMS OF GROWTH INHIBITION

Our hypothesis of growth regulation holds that as confluence is approached and passed, all the cells in a culture are gradually herded under growth inhibition, regardless of their individual circumstances. This results from contact-mediated transmission of inhibiting messages derived from a progressively rising proportion of cells not growing because their situation is unfavourable for some reason—for example, they are at the bottom of the sheet and starved. A "silent majority" of such unfavourably placed cells will emerge before the medium is exhausted; under standard conditions there will usually be some cells under restraint but otherwise capable of growth. On this hypothesis three factors are involved in cell-specific growth regulation:

(1) The definition of favourable compared to unfavourable circumstances that is used by the cells.

(2) The capacity of cells to make close contacts under the spread of conditions encountered under the culture regime employed.

(3) The capacity of cells to emit, transduce and respond to inhibiting messages.

Independent variations of these factors acting synergically or synergistically would provide a complex regulatory mechanism capable of a wide range of modulations.

With regard to the origin of inhibitory messages, Stoker (1967) has suggested that growth regulation could be achieved by the presence of a growth inhibitor molecule whose concentration built up as the cells became thick, an equilibrium being achieved between synthesis and loss. The role of this molecule could be to inhibit the entry of G_1 cells into S phase. In an appendix we consider quantitatively a model system whereby all cells produce an inhibitor molecule constitutively and lose a proportion of the inhibitory molecules by decay. In addition, those cells at the surface of the cell sheet may also lose their inhibitor to the medium. If the further assumption is made that the concentration of inhibitor is equilibrated throughout the culture by way of close contacts, then the mathematical analysis shows that some growth regulation may be achieved. Motility in the culture will render these junctions unstable, effectively isolating some cells from sources of inhibitor. These cells may then divide. As the culture becomes quiescent, intercellular contacts can be re-established and the inhibitor uniformly distributed.

The correctness of this approach is difficult to prove; however, the following experiment provides circumstantial evidence. The theory predicts that, provided the conditioned medium is not unduly depleted, growth will occur until the cells come into contact, at which point further division would be expected at a greatly reduced rate. New medium on the other hand should permit a faster rate of growth when the cells are touching, since motility would reduce the number of contacts mediating inhibitory influences.

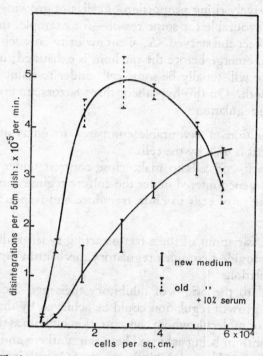

FIG. 3. The incorporation of tritium-labelled thymidine into adult skin fibroblasts at various densities in the presence of fresh and serum-supplemented conditioned media (confluent cells have a density of 8×10^4 cells per cm²).

Adult skin fibroblasts were inoculated at seven different densities ranging between 8×10^4 and $1 \cdot 15 \times 10^6$ per 50 mm dish in minimal Eagle's medium supplemented with $1 \cdot 5$ per cent foetal calf serum and antibiotics (this medium permits cell maintenance, but does not encourage growth). Nine dishes were set up in each class, and after 24 hours two were counted in a Coulter counter and one was fixed and stained. Of the remaining six, three were given fresh medium and three, conditioned medium supplemented with 10 per cent foetal calf serum, both sets of media containing 1μCi of tritiated thymidine per 5 ml. The conditioned medium was taken from confluent dishes on which it had lain for one or two days—

that is, long enough for motility factors to be reduced but before there is severe nutrient depletion. Twenty-four hours later the cells were fixed in 5 per cent TCA and the following day prepared for scintillation counting.

The results of the experiment are shown in Fig. 3. It can be seen that, in fresh medium, the counts are proportional to cell density at all but the extremes. The flattening-off at high densities represents the change to non-exponential growth. The curve showing the counts in conditioned medium is markedly different: at low density, there is a strong "conditioned medium" effect, and the counts are some ten times greater than those for new medium. The counts reach a peak at about one-third confluence but drop by a factor of two near confluence, to below the value for the fresh medium.

The experiment confirms the prediction that the change from exponential to non-exponential growth occurs at a lower cell density in conditioned medium than in fresh medium.

GROWTH OF CELLS ON COLLAGEN SUPPORTS

Practically all the work on growth regulation in cell cultures employs artificial glass and plastic substrates. It is also true that plastic and glass are not constituents of the body. Collagen, however, is, and when one starts growing cells on and in collagen supports, some surprising information on growth regulation comes to light. We make collagen substrates by first preparing a solution of collagen (from rat tail tendons) in a low ionic strength acid buffer, according to the method of Wood and Keech (1960). This is not pure collagen; Wood and Keech found about 10 per cent of the dry weight to be glucose residues. The ionic strength of an aliquot of collagen solution is raised to physiological levels by adding the appropriate amount of 10 × medium 199 and the pH is raised to 7·4 by adding 0·142 M-NaOH solution. The mixture is quickly transfixed to Petri dishes, spread and allowed to set. The result is a hydrated lattice of native collagen fibres containing about 0·13 per cent by weight of collagen. Fibroblasts attach and move happily on such a substratum, and they grow as fast as on plastic. It is difficult to determine whether they grow to the same stationary density on a collagen lattice as on plastic, because tensions in the dense cell sheets tend to tear the lattices. We have however counted $9·6 × 10^6$ cells on an intact lattice in a 50 mm dish. Collagen lattices can be applied as an overlay for cells plated on plastic. In this case, as long as the cells retain some attachment to the plastic they grow normally. So far there is nothing unusual to report. However, when fibroblasts are incorporated into a collagen lattice, by adding them to the mix, the cells grow if at all only very slowly, even though they spread out and elongate, using the lattice as a substratum. Under the same conditions HeLa cells and a line of SV40-transformed rat fibroblasts form colonies.

It is interesting to compare the behaviour of cells in relation to agar and collagen substrates. Normal cells will not grow on an agar-stiffened medium (unless excessively thick) nor in agar suspension. Transformed cells will grow in agar suspension. Normal cells grow *on* a collagen lattice, but not *in* a collagen lattice; transformed cells (judging from the SV40-transformed rat cells and some other cell lines also) will grow in collagen lattices. Non-growing normal cells in agar suspension remain unattached and rounded; non-growing normal cells in collagen lattices are attached and elongated.

In summary, our results show that:

(1) Normal fibroblasts grow well on a two-dimensional collagen lattice substratum.

(2) Normal fibroblasts grow at a greatly reduced rate, if at all, when embedded in a three-dimensional collagen lattice within which they attach and elongate normally.

These results lead on to the suggestion, worth further exploration, that the rapid growth of normal fibroblasts *in vitro* compared with that *in vivo* may result from the removal of the cells from their *in vivo* collagenous matrix.

APPENDIX

An essential of any theoretical model of growth regulation *in vitro* is the demonstration of some property of the culture which changes as cell number increases. A second requisite must be feedback from this property to some growth-linked parameter so that change in the property implies change in the probability of cell division. Following Stoker (1967; see main text, p. 191) we postulate the existence of an inhibitory molecule and suggest that its role is to lower the probability of the entry of G1-phase cells into S phase. The purpose of the exercise is to show that as the culture grows, so the concentration of the molecule grows, and regulation sets in. We assume that, in a culture of N cells,

(i) the molecule is produced constitutively by all cells at a rate p which is independent of its concentration C. The effect of the inhibitor on the growth rate is given by $f(C)$;

(ii) the molecule decays at a rate $q \times c$ either because it is inherently unstable or because it is degraded;

(iii) those cells at the surface of the culture (a number A equal to one monolayer equivalent and assumed to be independent of N) may also lose inhibitor to the medium at a rate $r \times C$. Inhibitor in the medium is assumed to be sufficiently diluted that reverse transport is negligible (cell: medium volume is $\sim 1:150$).

(iv) there is free interchange of the molecule between all cells so that its concentration is uniform in the culture. Moreover, the equilibration time is very much less than the division time. If interchange occurs through the tight junctions known to exist in fibroblasts in culture, the molecular weight of the molecule is probably less than about 1000 (Furshpan and Potter 1968).

The consequence of these conditions is that growth in the culture is described by the equations

$$N \frac{dC}{dt} = N.p - C[N.q + A.r]$$

$$\frac{dN}{dt} = \frac{N}{L} f(C)$$

$f(C)$ was chosen heuristically to be $1/1 + C^{5/2}$; experimentally, $A = 1 \cdot 5 \times 10^6$ and $L = 12$ hours (so that when $C = 1$, the doubling time is 24 hours). Growth was started at confluence and, as the equations are not sensitive to the initial value of C, this was set arbitrarily as 1. The equations are also insensitive to the absolute values of p, q and r; only their ratio matters. P was therefore set to 1 and the equations

$$\frac{dc}{dt} = p - C\left[q + \frac{A}{N}.r\right]$$

$$C_0 = 1$$
$$N_0 = 1 \cdot 5 . 10^6$$

$$\frac{dN}{dt} = \frac{N}{L} \frac{1}{1 + C^{5/2}}$$

were solved (on an IBM computer) for various values of q and r.

Figure 4 shows the best set of growth curves achieved (by setting $q = 0 \cdot 1$ and varying r). It is clear from the log plot that a degree of regulation may be achieved with maximum cell densities ranging from $1 \cdot 7 \times 10^6$ cells per 50 mm dish upwards. However, inhibition is not complete: these assumptions show that an initial growth rate will slow down by a factor of about two as the culture thickens and a lower proportion of total inhibitor escapes through the surface.

A complete proof of this model would require the isolation of the inhibitor molecule followed by an assay of its effectiveness *in vitro*. This is an extremely difficult enterprise partly because of the postulated instability of the molecule and also because its isolation and concentration, independent of toxic secretory products, pose great technical problems. Circumstantial evidence may be obtained by comparing growth rates when the cells are sparse and thick. It is unlikely, however, that the model is strictly

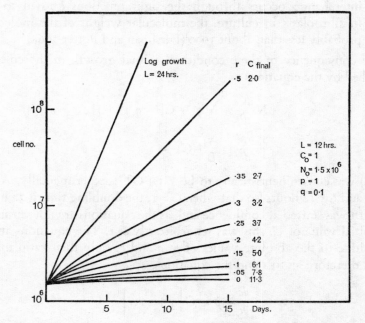

FIG. 4. Theoretical growth curves. (For detailed explanation, see text.)

correct: the simplicity of the mathematics is insufficient for an accurate description of the enormous biological complexity, and also the initial postulates are somewhat naive. We have excluded, for example, the possibility that the contact interactions may themselves affect the cell permeability parameters. However, a more accurate analysis would require many more arbitrary terms than the three employed here and hence would become little more than a curve-fitting procedure. The analysis does show that small variations in a single permeability parameter can produce large changes in doubling times and this, in turn, is suggestive in considering the very different growth rates that the same cell may show *in vivo* and *in vitro*.

SUMMARY

In investigating the growth regulating properties of human diploid fibroblasts we have examined their maximum cell densities under various conditions. The results suggest that those factors responsible for growth regulation are transmitted via intercellular contacts and not through the medium. Time-lapse photography and other data show that whereas fresh medium supports both motility and some growth in thick cultures, conditioned medium with added serum supports neither. A hypothesis to explain growth regulation by the action of a single molecular species inhibiting the entry of cells into S phase is put forward and its mathematical implications are examined (see Appendix). It has also been observed that whereas certain cell lines will grow both *in* and *on* collagen gels, early sub-culture fibroblasts will grow only *on* the gels.

REFERENCES

ELSDALE, T. R. (1968) *Expl Cell Res.* **51**, 439–450.
ELSDALE, T. R. and FOLEY, R. (1969) *J. Cell Biol.* **41**, 298–311.
FURSHPAN, E. J. and POTTER, D. D. (1968) *Curr. Top. Dev. Biol.* **3**, 95–127.
GRIFFITHS, J. B. (1970) *J. Cell Sci.* **6**, 739–749
SEFTON, B. M. and RUBIN, H. (1970) *Nature, Lond.* **227**, 843–845.
STOKER, M. (1967) *Curr. Top. Dev. Biol.* **2**, 107–128.
WOOD, G. C. and KEECH, M. K. (1960) *Biochem. J.* **75**, 588–597.

DISCUSSION

Montagnier: How pure is your collagen preparation?

Elsdale: The collagen is not pure. Wood and Keech (1960) find that about 10 per cent of the dry weight of this material is provided by glucose residues.

Montagnier: So the inhibitory effect may be due not to collagen but to sulphated polysaccharides contained in it. This would explain why your normal cells would not grow when suspended in the collagen gel.

Elsdale: Then why do they grow *on* the gel?

Montagnier: Do they spread on it?

Elsdale: Yes, and they spread out when they are in it, too.

Bard: The fibroblasts also make extremely strong adhesions to the collagen. You cannot trypsinize cells on collagen; you have to add collagenase to remove the collagen before the cells will round up.

Elsdale: Normal cells fail to clone in agar suspension because substrate requirements are unsatisfied, as evidenced by the failure of cells to stretch out normally. This is not true of cells within a collagen lattice; here the cells are beautifully stretched out. There is no reason to think that substrate requirements are not satisfied, yet the cells are largely inhibited in their growth. We do not know why.

Rubin: When one works at very high cell concentrations one has to worry about depletion of the medium. With medium 199 and chick cells, roughly 1 ml of medium supports the division of about a million cells; with 5 ml of medium we couldn't get very far above 5 million cells without frequent medium changes, and at 10 million cells the medium changes would have to be very frequent to avoid depletion. To what extent have you tested for the depletion of your medium as a possible explanation for some of your global phenomena?

Secondly, it was not clear to me why you thought there should be more cells when you added collagen. In the absence of collagen you had clumps of cells, and I wonder if you had calculated from the number, thickness and area of these clumps that you would expect to get more than about a 20 per cent difference from the cells plated on collagen?

Elsdale: I did not necessarily expect the difference; we did the experiment to see what would happen.

It is possible that the medium on dense cultures becomes depleted before it is changed. What is certain is that cells must enjoy a period of plenty after each change. The significant question is why, when a culture is dense, can the cells no longer employ these periods of plenty for nett growth?

Rubin: Certainly, but with little medium and lots of cells there would be enough nutrient for only a fraction of the population to divide before depletion became limiting, rather than the topological relationships of the cells. In order for us to get up to 17 million cells we had to change the medium twice a day for over a week.

Stoker: Am I right that in the first experiment, if it were not a global regulatory phenomenon you would expect a higher final density in the ridged sheet?

Elsdale: If growth were dependent on local density we would expect larger stationary populations in cultures where the cellular distribution was markedly uneven. This follows because ridges form by cell migration, resulting in a depopulation in the valleys between. On the local density hypothesis, growth should continue for longer in the valleys. However, we count the same number of cells in these cultures as in uniform cultures, a result that gives no support to the local density dependent inhibition hypothesis.

Abercrombie: We have estimated mitotic indices in different parts of single cultures of chick and guinea-pig fibroblasts (Abercrombie, Lamont and Stephenson 1968) and there is a negative correlation within one culture between local cell density and local mitotic rate; this is in regions of 2×10^4 cells per μm^2.

Rubin: Actually, Zetterberg and Auer (1970) showed the same thing with primary mouse kidney cultures.

Elsdale: I am quite prepared to believe that the mitoses may be predominantly in the thinner regions of a dense culture with ridges. These cultures however, do not grow to higher densities than uniform cultures.

Rubin: Once again, you must worry about medium depletion being the limiting factor when you have upwards of 10 million cells in a culture and are changing the medium every other day.

Elsdale: Medium removed from dense stationary cultures will support the growth of cells at lower densities. This test does not suggest a severe depletion of the medium. I am not convinced that it helps to consider growth regulation in terms of medium depletion when the medium is changed at regular intervals. The problem can be seen in these terms. Griffiths (1970) concludes that cells in dense cultures become inefficient utilizers of the medium. One could therefore envisage a situation where conditions for the growth of dense cells were satisfied for so short a period after each medium change that no nett growth resulted. Arguments can become circular, witness Yeh and Fisher (1969) who arrived at the notion of an inhibitor that only inhibited cells at high densities, thus begging the question. If we grant for the purposes of argument that cells in dense cultures suffer periods of impoverishment (regardless of the absolute amounts of nutrients present) it still has to be determined case by case whether this is the cause of growth inhibition or merely an incidental fact of life at higher densities. Now, in one of our experiments we have been able to contrive a situation whereby of two populations of the same cells growing in the same medium, one grew while the other was inhibited, thus demonstrating a dissociation of growth inhibition from the state of the medium. It's true that this situation was something of a special case because a second cell type was present. However it does suggest that the onus should be upon proponents of a medium depletion hypothesis to demonstrate for their systems the absence of other growth regulating mechanisms effective before nutrition becomes the limiting factor.

Rubin: The fact that medium from these dense cultures supports growth of cells at lower densities is certainly an important test. However, a careful quantitative comparison of growth rates must be made in the used and fresh media over a period of several days, since there is the risk of partial depletion in which limited division of a small number of cells might occur, but a large number would be restricted. The rate of uptake into the cell of a number of components—inorganic phosphate, glucose, uridine, thymidine, to name a few—is distinctly slower in crowded cultures, so that a medium partly depleted of glucose might support growth in a sparse but not a dense population.

Bard: But what is the difference between sparse and crowded cells? This *is* the question.

Rubin: Local pH! Surface area, motility and membrane potential are also different.

Taylor-Papadimitriou: You suggested, Dr Elsdale, that the burst of mitoses that you see on changing the medium follows from stimulation of movement. We made time-lapse films of the abortive transformation of cultures of BHK21 cells infected with polyoma virus and kept in medium containing 0·8 per cent of gamma-depleted serum. Control cultures of uninfected cells kept in gamma-depleted serum were also examined. We saw that although only the virus-infected cells went on to divide, the extent of movement was roughly the same in both infected and control cultures. The kind of movement observed, however, was quite different. The control uninfected cells moved very much in parallel, with cells going alongside each other, whereas the infected cells seemed to move randomly.

Stoker: It was a different type of movement. In the films which we studied we concluded that movement as such was not sufficient to initiate division; it had to be movement leading to loss of contact (Dulbecco and Stoker 1970).

Elsdale: Our idea here is that if one has inhibitory influences that are dependent upon, let us say, low-resistance junctions, anything that breaks these junctions will break down this inhibitory communication in the culture. Some of the more favourably placed cells, perhaps those at the top of the sheet, which were held down under inhibition because they were receiving messages previously, will then be allowed to go. We don't necessarily say that every cell that is moving goes, or that cell movement is directly affecting cell growth: it's breaking down communication in the culture.

Rubin: Several laboratories have looked at the extent to which communication breaks down in a culture after a change of medium (Furshpan and Potter 1968; Borek, Higashino and Loewenstein 1969). In some cases there was some breakdown but in others there was not. We have looked at it in multiplying cells and found no breaking of coupling.

Elsdale: But you said yourself that your chicken cells have no saturation density.

Rubin: That is not to say, however, that there is no effect of cell density on growth rate. I re-emphasize, the growth rate is markedly slowed at high density. I do want to get away from the all-or-none implications of the term "saturation density" which may apply strictly only to 3T3 cells, which are hardly a model for normality.

Todaro: Dr Mitchell Gale has been looking at sparse cultures (around 10^3 cells per dish) of 3T3 cells in gamma-depleted serum to which serum is then added. In gamma-depleted serum the cells barely move even though they are not in contact with other cells. When serum is added, within hours they begin to move in a random fashion and they go on to divide. So we are adding the serum factor(s) and stimulating cell movement. I don't see how it could be related to cell-to-cell communication in this situation. In these experiments it seems that the initiation of movement precedes cell division. What they have been trying to pin down is whether the increased rate of cell movement precedes the first round of cellular DNA synthesis.

Stoker: But this could be additional evidence for a point we reached earlier: that there seem to be two different mechanisms for initiation, one dependent on serum and the other, perhaps, on loss of topoinhibition.

The fact that you get movement and then DNA synthesis, independently of breaking contacts, could be due to the initiation of DNA synthesis by serum.

Todaro: There is a relation between movement and cell division. The SV40-transformed 3T3 cells in the same (gamma-depleted) medium move around quite extensively while the untransformed 3T3 cells do not do so until they have been stimulated by serum.

Weiss: Curtis and Greaves (1965) described a serum factor, also precipitable with the gamma globulin fraction, which inhibits cell aggregation and diminishes contact inhibition of movement. This factor might be responsible for initiating DNA synthesis. Martin Evans and I are pursuing this idea.

Dulbecco: We can add something further on this question of the correlation of movement and the activation of DNA synthesis from experiments with polyoma viruses, because the normal 3T3 cells don't move and do not synthesize DNA without serum but when they are infected by polyoma virus they start moving and making DNA. Using a temperature-sensitive virus mutant of group 5 the two changes can be dissociated, because DNA synthesis is very temperature dependent but movement is almost independent of temperature. So it would seem that the correlation is not complete. It could be that movement and DNA synthesis usually happen to occur at the same time because the same conditions stimulate both.

Elsdale: Clearly an association between motility and growth is interesting in a wider context than we have envisaged for the purposes of accounting for growth regulation in our system. The possibility of a direct relationship under certain circumstances is something we have not considered.

We are envisaging an indirect relationship only, whereby motility secondarily affects a primary relationship between cell-to-cell communication and growth regulation. We suggest, without specifying their nature, that cell-to-cell contacts, transducing inhibiting messages within a culture, are sensitive to cell motility. There will be fewer transducing contacts when the cells are motile, and more when the cells are immobile. In this scheme, motility directly affects the efficiency of message transduction, thereby indirectly affecting growth regulation.

Stoker: So far you are suggesting that movement is necessary but not enough; is there any example of initiation of cell DNA synthesis in the absence of initiation of movement? So far there isn't a virus mutant that does it.

Clarke: Could one consider two alternative models for such cells as 3T3 and Dr Elsdale's fibroblasts, which define a possible relationship between cell movement and DNA synthesis: (a) Resting cells in a confluent culture do not perform membrane functions which are associated with movement (? transport). (b) They are not able to react readily with serum and, *therefore*, are unable to perform these functions and are, *therefore*, unable to initiate DNA replication.

And alternatively: (a) Resting cells in confluent culture are able to react with serum but, due to some inhibition associated with their close proximity, they are unable to perform some membrane functions associated with movement. (b) They *therefore* cannot initiate DNA replication.

I think both Dr Abercrombie and Dr Holley have systems in which movement and growth are dissociated, so one might need to postulate that the membrane activity is usually, though not necessarily, associated with cell movement. I think these alternatives are testable.

Elsdale: Fresh serum added to a culture in old medium causes only a little cell division and virtually no resurgence of motility in dense cultures. According to Griffiths (1970) the growth induced can be accounted for by serum amino acids. It appears therefore that in the presence of old medium, non-amino acid serum components stimulate neither growth nor motility.

Rubin: Adding serum to density-inhibited chick cell cultures gives a DNA synthesis response which cannot be duplicated by adding a combination of amino acids to the medium. Was it not when you reached very high cell densities that you began to get a failure of response?

Elsdale: No, it was a fairly low density. The highest density we used was just over a million cells in a dish; the cells were only approaching confluence.

Todaro: With 3T3 cells we found that it doesn't matter; you can add serum to the old medium and that will stimulate cell division almost as efficiently as if the cells are put into fresh medium. Dialysing the old medium did not restore activity (Todaro, Lazar and Green 1965).

Bürk: This is true of BHK cells as well.

Elsdale: Griffiths is working with WI 38 cells which are fairly similar to ours.

Dulbecco: I would like to bring in a result obtained in a different system, in epithelial BSC-1 cells from African green monkeys. These cells were grown to a confluent layer, wounded, transferred to medium without serum and after 28 hours were given a pulse of [³H]thymidine and radio-autographed. Far from the wound there is essentially no DNA synthesis; in the wound there is mass migration which generates a new edge, but the old edge is still identifiable by piled-up cells. The area where practically all cells are making DNA extends into the cell layer quite a way beyond the original wound edge. This seems to imply that in making a wound, one perturbs even cells which don't actually migrate; they might have moved slightly but their reciprocal relationships are largely unchanged, so the part played by movement in this case must have been very small.

Rubin: In the guinea-pig ear, the peak numbers of [³H]thymidine-labelled nuclei occur 300–400 cells away from the edge of the wound (Hell and Cruickshank 1963). In the rat, the initial reaction is claimed to be 500 basal cells from the edge of the wound (Block, Seiter and Oehlert 1963).

Stoker: Loewenstein measured coupling between cells in series extending back from a wound in urodele skin and found that uncoupling extended several cells back from the wound edge for a short period (Loewenstein and Penn 1967).

Dulbecco: These results seem to suggest that some local perturbation in the relation between the cells, perhaps a transient one, is enough to produce activation of DNA synthesis.

Rubin: That is apparently true in the wounded lens, where a stimulus to mitosis is propagated over the surface at the rate of 17 μm per hour (Harding and Srinivasan 1961).

Weiss: What they didn't say, but it is very clear in their pictures, is that the cell contact became very loosened in front of the labelling band. They also got several rings or generations of mitosis.

Abercrombie: It is a common finding in epithelial wound-healing that the cells which actually move into the wound do not divide; mitosis is often confined to the cells behind.

Clarke: BSC-1 cells require serum for growth, but the cells in wounds made in resting layers do not need serum for DNA synthesis (Dulbecco 1970). Could not these cells have reacted with serum in dense culture and need only "space" to perform the function for which the serum has provided the information? This then leads to movement and to initiation of DNA replication.

Dulbecco: In any case the movement was mass movement; perhaps the relationships between cells do not undergo important changes in this movement. Are tight junctions disturbed during this movement of epithelial cells?

Abercrombie: This is not known. Probably the junctions are not disturbed.

Warren: Lane and Becker (1966) found that one of the first changes when you remove most of the liver of a rat is that in the remaining liver there is an increased intercellular distance or space; processes from each cell that interdigitate now disengage, pull apart, and the space is increased. This would seem to precede a terrific burst of DNA synthesis and all the other changes that go on in regenerating liver. Instances have been described where there is sudden movement, a sort of wriggling, one cell trying to get away from another, before a burst of DNA synthesis occurs. Perhaps one of the critical things in order for cells to commence or continue to divide is for them to isolate themselves from one another physically; so perhaps one of the critical factors is intercellular distance. Is it possible that in a tumour there is a permanently increased intercellular distance? I don't mean an obvious one that can readily be seen in the electron microscope. In the case of regenerating liver you are essentially creating a tumour, temporarily. Since the liver cells have the capacity to come together again and eliminate the increased intercellular space the tumour situation is eliminated. If the increased intercellular space could not be eliminated, perhaps due to high turnover, a tumour would result.

Abercrombie: Loewenstein and Kanno (1967) showed that in regenerating liver the coupling between cells was kept.

Rubin: They measured it at 20, 48 and 72 hours and found no uncoupling, but in skin Loewenstein had to look within a few minutes after wounding to find uncoupling (Loewenstein and Penn 1967).

Elsdale: Dr Warren's comments are wholly relevant to what we have in mind. We believe that motility promotes a transient isolation of the cells by causing a reduction in those contacts by which cells pass around inhibiting messages. Note, however, that in stationary cultures few cells are actually induced to divide. One could perhaps relate the stimulation of growth in this situation to the local stimulation of growth induced by wounding, even although those situations are superficially dissimilar. The underlying physiological situation may be the same in both cases, the operative factor being isolation of the cells from inhibiting messages. Effective isolation for the purposes of escape from growth inhibition may not necessarily require a gross physical separation as observed in wounding experiments when cells migrate into the wound space.

Stoker: I was impressed by the hole-in-the-middle experiment, but this seems to me to say that it is *not* a global phenomenon, otherwise the cells in the middle ought to obey the lower saturation density rule, or maybe something half way. What happens at the edge?

Elsdale: There is a pretty sharp cut off at the edge.

Stoker: Can you explain this?

Elsdale: I should have mentioned an important point about the procedure in these experiments. After the second inoculation of cells, the central portion of the culture is kept isolated from the surrounding zone by scraping a thin circle around the central region. Thus isolated, no interaction between the two regions can occur except through the medium, and this interaction appears to be insignificant.

Regarding the ratio of low density to high density cells required to inhibit the HD cells, this has been investigated by plating dishes with equal numbers of cells in different proportions. Our results show that LD cells must outnumber HD cells several times before significant inhibition of HD cells is obtained.

Preliminary results suggest that when our hole-in-the-middle experiment is done without further scraping after the inoculation with HD cells, permitting the fusion of the two zones across their boundary, roughly the same result as before is obtained. A possible explanation is provided by the above experiments on the growth of mixed platings. There will be no inhibition within the central region in the hole-in-the-middle situation until the HD cells growing there have achieved confluence; at that stage the stationary LD cells surrounding the HD cells have a density only two or three times higher, whereas the mixed-cell experiments show that a difference of five to six-fold is required to achieve good inhibition of HD cells.

Weiss: We get essentially the same results in experiments on chick embryo cells with 3T3 cells or polyoma-transformed BHK21 cells with mouse cells (Weiss and Njeuma 1971). With a mixture of cells the high density cells are only inhibited by an excess of low density cells and even then, if polyoma cells are in contact with each other, they will proliferate regardless of the mouse cells.

Stoker: What about the 3T3 mixed-cell experiments with the different saturation densities?

Todaro: The mixtures tend to grow to the lower density. If you mix a reasonable excess of 3T3 with 3T12, with saturation densities of one million and about ten million, you obtain a final saturation density of 1·5 or 2 million, which is much closer to that of the 3T3 cell. One can pick out discrete areas on the plate containing the two cell types. We get the same result with human embryo fibroblasts mixed with 3T3 cells. The final density is nearer the 3T3 saturation density.

REFERENCES

ABERCROMBIE, M., LAMONT, D. M. and STEPHENSON, E. M. (1968) *Proc. R. Soc. B* **170**, 349–360.
BLOCK, P., SEITER, I. and OEHLERT, W. (1963) *Expl Cell Res.* **30**, 311–321.
BOREK, C., HIGASHINO, S. and LOEWENSTEIN, W. R. (1969) *J. Membrane Biol.* **1**, 274–293.
CURTIS, A. S. G. and GREAVES, M. F. (1965) *J. Embryol. exp. Morphol.* **13**, 309–326.
DULBECCO, R. (1970) *Nature, Lond.* **227**, 802–806.
DULBECCO, R. and STOKER, M. G. P. (1970) *Proc. natn. Acad. Sci. U.S.A.* **66**, 204.
FURSHPAN, E. J. and POTTER, D. D. (1968) *Curr. Top. Dev. Biol.* **3**, 95.
GRIFFITHS, J. B. (1970) *J. Cell Sci.* **6**, 739–749.
HARDING, C. V. and SRINIVASAN, B. D. (1961) *Expl Cell Res.* **25**, 326–340.
HELL, E. A. and CRUICKSHANK, C. D. (1963) *Expl Cell Res.* **32**, 354–357.
LANE, B. P. and BECKER, F. F. (1966) *Am. J. Path.* **48**, 183–196.
LOEWENSTEIN, W. R. and KANNO, Y. (1967) *J. Cell Biol.* **33**, 225–233.
LOEWENSTEIN, W. R. and PENN, R. D. (1967) *J. Cell Biol.* **33**, 235–242.
TODARO, G. J., LAZAR, G. and GREEN, H. (1965) *J. cell. comp. Physiol.* **66**, 325.
WEISS, R. A. and NJEUMA, D. (1971) This volume, pp. 169–184.
WOOD, G. C. and KEECH, M. K. (1960) *Biochem. J.* **75**, 588–597.
YEH, J. and FISHER, H. W. (1969) *J. Cell Biol.* **40**, 382–388.
ZETTERBERG, A. and AUER, G. (1970) *Expl Cell Res.* **62**, 262–270.

TRANSPORT AND PHOSPHOLIPID CHANGES RELATED TO CONTROL OF GROWTH IN 3T3 CELLS

Dennis D. Cunningham* and Arthur B. Pardee

Department of Biochemical Sciences, Princeton University, Princeton, New Jersey

The growth of 3T3, and of most "normal" cells in culture, is strongly dependent on cell density and serum levels. These cells stop dividing after growth to confluency when they are grown in the usual medium containing 10 per cent calf serum (Todaro and Green 1963; Todaro, Green and Goldberg 1964). The final saturation density of the cells, however, is proportional to the amount of serum in the medium (Todaro *et al.* 1967; Holley and Kiernan 1968). In contrast, growth of virally transformed 3T3 cells is not dependent on cell density (Todaro, Green and Goldberg 1964), and they have a reduced requirement for serum growth factors (Todaro *et al.* 1967; Holley and Kiernan 1968).

This system seemed well-suited for studies on whether changes in the cell membranes might be related to control of growth. Our results show that there are specific transport and phospholipid changes related to density- and serum-dependent growth control in 3T3 cells. The same changes do not take place in polyoma virus-transformed 3T3 (Py3T3) cells. These results are consistent with the hypothesis that specific transport changes might participate in growth regulating events (Pardee 1964).

MEMBRANE CHANGES FOLLOWING INITIATION OF CELL DIVISION

Cell division can be initiated in confluent 3T3 cells by adding fresh serum (Todaro, Lazar and Green 1965). This treatment activates a fraction or all of the "contact inhibited" cells, depending on the amount added. Synchronous DNA synthesis reaches a peak at about 20 to 22 hours, and mitotic cells are most frequent about 30 hours after adding fresh serum. In addition, increased incorporation of uridine into RNA is evident within 10 minutes, and a wave of increased incorporation of amino acids

* *Present address:* Environmental Interactions Curriculum, California College of Medicine, University of California, Irvine, California 92664.

into protein occurs several hours later (Todaro, Lazar and Green 1965; Todaro et al. 1967).

Our first evidence that membrane changes might be related to this initiation of division in 3T3 cells came from experiments on the incorporation of [^{32}P]phosphate into total phospholipids. Treatment of confluent cells with fresh serum brought about a several-fold increase in this incorporation within 5 minutes (Cunningham 1969). Initiation of division of human lymphocytes by phytohaemagglutinin brings about a selective increase in the incorporation of [^{32}P]phosphate into phosphatidyl inositol

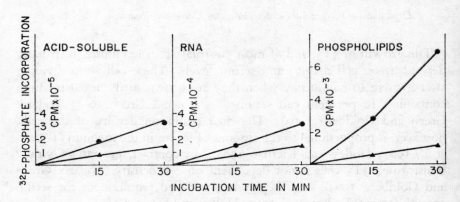

FIG. 1. Effect of addition of serum on the incorporation of [^{32}P]phosphate into the acid-soluble, RNA and phospholipid fractions of confluent 3T3 cells. Measurements were made three days after a medium change on control (triangles) and serum-treated cells (circles). The time is minutes after the addition of [^{32}P]phosphate and dialysed fresh calf serum to a final concentration of 10 per cent. (From Cunningham and Pardee 1969, with permission.)

(Fisher and Mueller 1968). To see if the increase in 3T3 cells was also selective, ^{32}P-labelled phospholipids from control and serum-treated confluent cells were separated by chromatography and counted. In contrast to lymphocytes, incorporation into all the major phospholipids was stimulated to the same extent in 3T3 cells. This suggested that the increased incorporation might be a result of increased phosphate transport brought about by the fresh serum. We checked this by examining the uptake of phosphate into the acid-soluble fraction of control and serum-treated confluent 3T3 cells, and found that fresh serum indeed brought about a rapid increase of phosphate transport, as shown in Fig. 1 (Cunningham and Pardee 1969). The effect, however, was not large enough to account completely for the increased labelling of the phospholipids (Fig. 1). Initiation of division in confluent 3T3 cells by fresh serum, therefore, was accompanied by two very early membrane changes: increased synthesis or turnover of all the major phospholipids, and increased phosphate transport.

We also examined the incorporation of [³²P]phosphate into RNA of control and serum-treated confluent 3T3 cells, and found that the increase in phosphate uptake was equal to the increase in incorporation of phosphate into RNA (Fig. 1). This suggested that the rapid stimulation of RNA labelling brought about by fresh serum was not a result of increased synthesis (Todaro, Lazar and Green 1965; Todaro *et al.* 1967) but instead a result of a higher specific activity of the precursor pool caused by increased phosphate transport. We checked this using other RNA precursors. Fresh serum also brought about a rapid and large increase in uridine

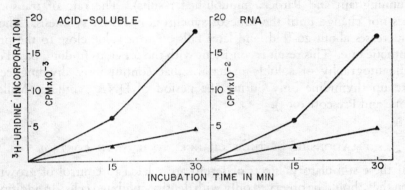

FIG. 2. Effect of addition of serum on the incorporation of [³H]uridine into the acid-soluble and RNA fractions of confluent 3T3 cells. The experiment was performed on control (triangles) and serum-treated (circles) cells, as described in Fig. 1. (From Cunningham and Pardee 1969, with permission.)

transport. Moreover, the increase was equal in magnitude to the increased incorporation into RNA (Fig. 2). Results almost identical to those with uridine were obtained with cytidine and guanosine. Adenosine transport, however, was quite different. The addition of fresh serum had no effect on the transport of adenosine by confluent 3T3 cells. Moreover, it had no effect on the incorporation of this precursor into RNA during the first 30 minutes. We also measured the specific activities of the phosphate- and uridine-containing acid-soluble pools in control and serum-treated confluent 3T3 cells, and found that the increases brought about by fresh serum were equal to the increases in RNA labelling. These results demonstrated that the early increased incorporation of certain precursors into RNA of confluent 3T3 cells is a result of increased transport brought about by fresh serum (Cunningham and Pardee 1969).

The early stimulation of transport is specific, since no serum effect was observed for adenosine, 3-O-methyl-D-glucose or amino acids. In addition, it appears that the stimulation is brought about by specific serum components rather than by a non-specific effect of protein. Fractionation of dialysed fresh serum on Sephadex G-200 demonstrated that the activity which stimulated phosphate transport was associated primarily with the albumin fraction. Checks of crystalline serum albumin revealed significant activity in only three out of seven different lots tested. The stimulatory material, therefore, is probably similar in size to albumin or else tightly bound to it (Cunningham and Pardee 1969).

Transport of thymidine into the acid-soluble fraction is also modified during the proliferative response in serum-treated confluent 3T3 cells (Cunningham and Pardee, unpublished results). The rate of transport does not change until the time of synchronous DNA synthesis. Then, it increases about 20-fold, and later falls off to a value close to the pre-synthetic rate. This result is consistent with the previous finding, based on radioautography of soluble materials, that continuously dividing cells take up thymidine only during the period of DNA synthesis (Miller, Stone and Prescott 1964).

RELATIONSHIP OF THESE CHANGES TO GROWTH CONTROL

If these stimulatory effects of serum are related to control of growth, then they should be observed only with density-inhibited cells. In addition, we should expect rates for non-confluent 3T3 cells to be higher than confluent 3T3 cells, whereas rates for Py3T3 cells should not decrease after growth to confluency. We tested this for uridine transport by measuring the rates of uptake into the acid-soluble fraction of non-confluent and confluent 3T3 and Py3T3 cells during a 15-minute labelling period. These measurements were standardized per mg of cell protein (Foster and Pardee 1969) and are shown by the plain bars in Fig. 3. The rate of uptake by confluent 3T3 cells was five- to six-fold lower than by non-confluent 3T3 cells. In contrast, uptake by Py3T3 cells was reduced only about 25 per cent after the cells became confluent. The hatched bars in Fig. 3 show rates of uridine uptake during the 15-minute interval immediately following addition of fresh serum to the cells. A large increase was observed for confluent 3T3 cells, but not for the rapidly growing Py3T3 or non-confluent 3T3 cells (Cunningham and Pardee 1969). These results suggest that stimulation of uridine transport is related to the initiation of cell division in confluent 3T3 cells. As with uridine, the rate of phosphate uptake decreased about three-fold after 3T3 cells became confluent. It increased about 50 per cent after Py3T3 cells became confluent. Like uridine, addition of fresh serum brought about a large increase in phosphate uptake only by confluent 3T3 cells (Cunningham and Pardee 1969).

Specific transport inhibitors released by 3T3 cells participate in the inhibition of transport by confluent 3T3 cells (Pariser and Cunningham 1971). These inhibitors do not affect transport by Py3T3 cells. They are dialysable, and the one for uridine appears identical to the inhibitor reported by Yeh and Fisher (1969) which reduces the serum-stimulated incorporation of uridine into RNA by confluent 3T3 cells.

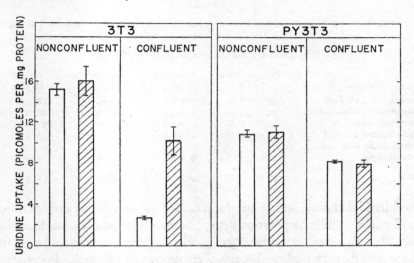

FIG. 3. Uptake of [³H]uridine into the acid-soluble fraction of confluent and non-confluent 3T3 and Py3T3 cells during a 15-minute incubation period three days after a medium change (plain bars). The hatched bars show uptake into cells to which fresh serum was added (final concentration of 10 per cent) immediately before incubation. Vertical lines show ± one standard deviation from the mean. (From Cunningham and Pardee 1969, with permission.)

The rapid stimulation of the synthesis or turnover of phospholipids brought about by fresh serum occurred only with confluent 3T3 cells. Py3T3 and non-confluent 3T3 cells showed much smaller increases.

DIFFERENCES IN PHOSPHOLIPIDS BETWEEN NON–CONFLUENT AND CONFLUENT 3T3 CELLS

To see if other membrane changes might be related to contact inhibition of growth, we examined the incorporation of [³²P]phosphate into individual phospholipids (Cunningham 1970). Non-confluent and confluent 3T3 cells were labelled for two hours with [³²P]phosphate, and labelled phospholipids were extracted and separated by chromatography. The amount of radioactivity in each phospholipid was then expressed as a percentage of the total labelled phospholipids (Table I). As shown, the percentage incorporated into phosphatidyl choline more than doubled after 3T3 cells became confluent. In contrast, it decreased about five-fold for phosphatidyl ethanolamine. To see if these changes were related to density-inhibition of growth, identical experiments were carried out on Py3T3 cells. As shown in Table I, the percentage of radioactivity incorporated into

TABLE I

INCORPORATION OF [^{32}P]PHOSPHATE INTO INDIVIDUAL PHOSPHOLIPIDS DURING A TWO-HOUR PULSE

(Expressed as percentage of total phospholipids)

Phospholipid	Non-confluent 3T3	Confluent 3T3	Non-confluent Py3T3	Confluent Py3T3
Phosphatidyl choline	27·0±3	63·0±6	15·0±3	18·0±2
Phosphatidyl ethanolamine	24·0±4	4·8±0·4	27·0±3	18·0±3
Phosphatidyl serine	42·0±6	29·0±3	45·0±7	57·0±7
Phosphatidyl inositol	3·6±0·6	1·7±0·4	6·0±1	3·0±1
Phosphatidic acid and cardiolipin	2·4±1	0·6±0·2	5·3±2	2·3±1
Total	99·0	99·1	98·3	98·3

phosphatidyl choline was about the same for non-confluent and confluent Py3T3 cells. The incorporation into phosphatidyl ethanolamine decreased only about 30 per cent after Py3T3 cells became confluent. Thus, the changes in these two phospholipids were unique to 3T3 cells. In contrast, changes in phosphatidyl inositol and phosphatidic acid plus cardiolipin after growth to confluency were similar for both 3T3 and Py3T3 cells.

Do these labelling differences reflect different rates of net synthesis that would lead to a difference in the phospholipid composition of the cells, or are they a result of different rates of turnover? This question was answered by determining the phospholipid composition of non-confluent and confluent 3T3 and Py3T3 cells. These measurements were made by an isotope equilibration technique because of the difficulty of growing large numbers of non-confluent cells. The cell phospholipids equilibrate with the [^{32}P]phosphate in the medium within three days. Then, by determining the amount of radioactivity in each phospholipid and knowing the specific activity of [^{32}P]phosphate in the medium, it is possible to calculate the chemical amount of each phospholipid. These data are shown in Table II. As before, the value for each phospholipid is expressed as a percentage of the total phospholipids. It is apparent that the phospholipid compositions of non-confluent and confluent 3T3 and Py3T3 cells are about the same. Thus, the phospholipid labelling differences detected by two-hour pulses are primarily a result of different rates of turnover of the individual phospholipids.

It should be pointed out that all these studies were made on phospholipids extracted from whole cells. At this time, then, we do not know whether the rapid changes following the addition of serum, or the differences between non-confluent and confluent 3T3 cells, are restricted to specific membrane components of the cells. Fractionation of the cells before extraction of the phospholipids should allow us to answer this question.

TABLE II

PHOSPHOLIPID COMPOSITION

(Expressed as percentage of total phospholipids)

Phospholipid	Non-confluent 3T3	Confluent 3T3	Non-confluent Py3T3	Confluent Py3T3
Phosphatidyl choline	52·0	54·0	52·0	53·0
Phosphatidyl ethanolamine	27·0	29·0	25·0	26·0
Phosphatidyl serine	6·0	7·0	7·0	6·0
Phosphatidyl inositol	3·1	3·2	3·4	3·9
Sphingomyelin	7·0	3·0	6·0	7·0
Lysophosphatidyl choline	2·2	1·7	1·8	1·7
Phosphatidic acid and cardiolipin	2·7	2·1	4·8	2·4
Total	100·0	100·0	100·0	100·0

The transport and phospholipid changes described here appear to be related to control of growth in 3T3 cells, since they do not occur in Py3T3 cells. This kind of relationship, of course, does not necessarily mean that they are directly involved in a growth-controlling mechanism. In fact, it might appear that they are simply caused by different growth rates. However, the serum-induced responses can be detected within five minutes, and this strongly suggests that they are not consequences of altered growth rates, but instead are primary effects of the action of serum. In addition, the decrease in transport of phosphate and uridine after growth of 3T3 cells to confluency is a specific change and not simply a result of a slower metabolic state, since active transport of amino acids decreases only slightly after 3T3 cells become confluent (Foster and Pardee 1969).

SUMMARY

Specific and rapid membrane changes take place when cell division is initiated in "contact inhibited" 3T3 cells by adding fresh serum. This treatment brings about increased uptake of phosphate and uridine into the acid-soluble fraction as well as increased synthesis or turnover of all the major phospholipids. These changes can be detected within five minutes, and appear, therefore, to be primary effects of serum rather than consequences of altered growth rates. Transport of thymidine into the acid-soluble fraction also increases after the addition of serum to confluent 3T3 cells. A 20-fold increase occurs specifically at the time of synchronous DNA synthesis.

Inhibition of division following growth to confluency is also accompanied by specific membrane changes. Phosphate and uridine transport are much lower in confluent than non-confluent 3T3 cells. Specific transport inhibitors released into the medium participate in these decreases. In addition, after 3T3 cells grow to confluency, the rate of turnover of phosphatidyl choline relative to the other phospholipids more than doubles, while the rate for phosphatidyl ethanolamine decreases five-fold.

These membrane changes appear to be related to density- and serum-dependent growth control in 3T3 cells, since the same changes do not take place with 3T3 cells transformed by polyoma virus.

Acknowledgements

This work was supported by grants from the United States Public Health Service (AI-04409), American Cancer Society (E-555) and United States Atomic Energy Commission (Contract #AT (30-1)4108). We thank Miss Etheria Robinson and Miss Clara Potts for technical help. D.D.C. was supported by a Postdoctoral Research Fellowship from the National Science Foundation.

REFERENCES

CUNNINGHAM, D. D. (1969) *Fedn Proc. Fedn Am. Socs exp. Biol.* **28(2)**, 845, abst. 3281.
CUNNINGHAM, D. D. (1970) *Fedn Proc. Fedn Am. Socs exp. Biol.* **29(2)**, 605, abst. 2015.
CUNNINGHAM, D. D. and PARDEE, A. B. (1969) *Proc. natn. Acad. Sci. U.S.A.* **64**, 1049–1056.
FISHER, D. B. and MUELLER, G. C. (1968) *Proc. natn. Acad. Sci. U.S.A.* **60**, 1396–1402.
FOSTER, D. O. and PARDEE, A. B. (1969) *J. biol. Chem.* **244**, 2675–2681.
HOLLEY, R. W. and KIERNAN, J. A. (1968) *Proc. natn. Acad. Sci. U.S.A.* **60**, 300–304.
MILLER, O. L., STONE, G. E. and PRESCOTT, D. M. (1964) *J. Cell Biol.* **23**, 654–658.
PARDEE, A. B. (1964) *Natn. Cancer Inst. Monogr.* **14**, 7–20.
PARISER, R. J. and CUNNINGHAM, D. D. (1971) *J. Cell Biol.* in press.
TODARO, G. J. and GREEN, H. (1963) *J. Cell Biol.* **17**, 299–313.
TODARO, G. J., GREEN, H. and GOLDBERG, B. D. (1964) *Proc. natn. Acad. Sci. U.S.A.* **51**, 66–73.
TODARO, G. J., LAZAR, G. K. and GREEN, H. (1965) *J. cell. comp. Physiol.* **66**, 325–333.
TODARO, G. J., MATSUYA, Y., BLOOM, S., ROBBINS, A. and GREEN, H. (1967) In *Growth Regulating Substances for Animal Cells in Culture*, pp. 87–101, ed. Defendi, V. and Stoker, M. Philadelphia: Wistar Institute Press.
YEH, J. and FISHER, H. W. (1969) *J. Cell Biol.* **40**, 382–388.

DISCUSSION

Rubin: You have evidence of an inhibitor of transport in your used medium. Is the degree of inhibition of transport correlated with the degree of growth inhibition?

Cunningham: If medium which has supported the growth of 3T3 cells to confluency is added to *non-confluent* 3T3 cells, transport of phosphate decreases only slightly. The inhibitor for phosphate transport apparently is active only with confluent 3T3 cells. Transport of uridine by the non-confluent cells decreases about three-fold. I should point out that these measurements were made during the 20-minute interval immediately following the addition of the used medium to growing 3T3 cells, and it might be that larger decreases would be observed at a later time. If you now measure growth of non-confluent 3T3 cells in the same kind of used medium, the decrease in growth rate over a two-day period is at least five-fold.

Todaro: If we add 10 per cent serum to 3T3 cultures, or if we change the medium completely, both induce DNA synthesis and cell division but it will be slightly more efficient if the medium is changed. With 10 per cent serum approximately 6 per cent of the cells go on to divide in fresh medium, whereas after adding 10 per cent serum to the old medium, 3 per cent divide. The dose–response curve in this range is very steep, so that with 30 per cent serum or the equivalent of 100 per cent serum, practically all the cells can be induced to divide. We felt that the major factor was the addition of the serum, but that there was possibly some damping effect from the old medium, which could be due to inhibitors or to depletion of nutrients.

Rubin: If the altered transport of uridine and phosphate has any specific significance in regulating growth, it must be the phosphate which is limiting, since the cells synthesize uridine. Since it is unlikely that phosphate is limiting, the altered transport of these materials must reflect a more general change, such as that indicated by altered trans-membrane potential.

Clarke: Can insulin replace serum for any of these functions?

Cunningham: We have not tested the ability of insulin to bring about the rapid changes in uridine and phosphate transport.

Bergel: When you expose your cells to labelled inorganic phosphate or uridine or other compounds, do you take fresh cells each time or do you expose cells to labelled inorganic phosphate and uridine simultaneously, and measure any transport changes in the same cells?

Cunningham: In a few cases we measured the simultaneous uptake of [32P]phosphate and [3H]uridine into the same cells. But usually we measured the uptake of these compounds in separate cultures. The result is the same in either case.

Bergel: Does this mean that the compounds are not competing with each other when they are taken up, because one can assume that the cell—intracellularly—will phosphorylate uridine very rapidly and then has a certain requirement for inorganic phosphate?

Cunningham: Yes, uridine is phosphorylated very rapidly after it is taken up by the cells. But the pool size of inorganic phosphate is large in comparison with the amount of uridine taken up in these experiments.

Montagnier: Have you looked at the release of dialysable material into the medium of confluent and non-confluent cells, in order to obtain an idea of the kind of inhibitor it is?

Cunningham: We know very little about the nature of this inhibitor except that it is a material of low molecular weight, and we know that it is not uridine and not phosphate.

Taylor-Papadimitriou: You said that the uptake of thymidine was 20 times greater when the cells were in DNA synthesis. Did you have all the cells in DNA synthesis and were they confluent?

Cunningham: We started with confluent cultures of 3T3 cells, added fresh serum to a final concentration of 10 per cent, and then measured uptake of thymidine into the acid-soluble fraction at various times. The uptake did not change until the time of DNA synthesis. At 22 hours it had increased 20-fold. By 35 hours it had decreased to a value close to the presynthetic rate. The period of increased thymidine transport corresponded fairly closely in time to the period of DNA synthesis.

When fresh serum is added to a final concentration of 10 per cent, only about 5 per cent of the cells replicate their DNA and go on to divide. If more serum is added, more of the cells are activated, and it is likely that the increase in thymidine uptake would be greater.

Rubin: You now attribute the apparent increase in the synthesis of RNA to increased uptake of uridine. Do you think there is any difference in the rates of RNA synthesis in the inhibited cells and the stimulated cells? Can you account for all the difference in apparent incorporation into RNA by increased uptake? We also find that growing cells take up uridine faster than density-inhibited cells (Weber and Rubin 1971) but there is greater RNA synthesis as well.

Cunningham: It is now clear that the increased incorporation of certain precursors into RNA during the first 30 minutes after adding fresh serum to confluent 3T3 cells can be accounted for by increased transport of the precursors. The best evidence for this is that serum treatment brings about no change in either the rate of adenosine transport, or its rate of incorporation into RNA during this 30-minute interval. We haven't looked at times after this.

Todaro: When we added serum to confluent cells three days after a medium change we regularly found a burst of mitosis with a peak (about 6 per cent of the cells) at 30 hours, against a background of under 0·1 per cent mitosis in the controls (no serum addition). We then found a peak of thymidine incorporated into TCA-precipitable counts, by pulsing with tritiated thymidine at various times after adding the serum factor. By radioautography, under 0·1 per cent of the cells were positive until about 16 hours, when positive nuclei began to appear. With 10 per cent serum, 5–6 per cent of the cells were positive and about the same number would go into mitosis. Thus the great majority of the cells as a result of getting serum neither made DNA nor divided, and all the divisions were labelled. So we think that all the cells that make DNA at 16 to 24 hours go on to divide at 30 hours. The whole population is blocked in the G1 phase of the cell cycle. Looking earlier, we had found by uridine incorporation into TCA-precipitable counts a spike (about a 15-fold increase) at 30 minutes; then it came down and then went up again at 2 hours and drifted down. There was no increased incorporation of labelled proline or leucine into protein for the first hour or two and then a gradual rise reaching a peak at 2–4 hours. If we accept that the initial spike of uridine uptake is not related to new RNA synthesis, how far back can this be taken? What do we know about the events before the onset of DNA synthesis, which doesn't begin until about 16 hours? Do you think that the thymidine incorporation is also not directly related to the cell division at 30 hours?

Cunningham: If enough fresh serum is added to confluent 3T3 cells to cause all of them to divide, it can be demonstrated by chemical methods that there is a doubling of DNA between 15 and 30 hours (A. J. Levine, personal communication). Therefore, the large increase in [³H]thymidine incorporation into acid-insoluble material which peaks at about 20 to 22 hours indeed reflects increased DNA synthesis. What our results show is that during this period of DNA synthesis there is also an increase in the transport of thymidine into the acid-soluble fraction of the cells.

On the other hand, the early increased incorporation of [³H]uridine into RNA is not a result of an increased rate of RNA synthesis, but instead an increased rate of uridine uptake into the acid-soluble fraction. You mentioned that this increase reaches a peak at about 30 minutes after adding fresh serum and then tapers off. When we measured uridine transport using short five-minute pulses we also found that transport peaked at about 30 minutes and then decreased.

Holland: What exactly is the correlation which you are proposing exists between surface changes and DNA synthesis? Are you suggesting that the changes in permeability are somehow triggering replication?

Cunningham: It is probably true that a large number of events are involved in "triggering" replication. At present, we have no evidence that these transport changes are directly involved in starting up DNA replication. It is worth noting, however, that added thymidine can shorten an artificially extended G1 back to its minimum value (Tobey, Anderson and Petersen 1967). Changes in the pool size of thymidine-containing compounds, therefore, might have something to do with the timing of DNA synthesis. This possibility is currently being studied.

It is possible that the increased thymidine transport is a *result* of increased rates of DNA synthesis, which could deplete thymidine pools and result in increased uptake of thymidine. The increased transport might be a result of increased levels of thymidine kinase during S. Experiments using inhibitors of DNA synthesis and 3T3 cells lacking thymidine kinase should shed some light on events which bring about the increase in thymidine transport.

Holland: In bacteria there is no evidence that the precursor pools normally influence the initiation or the rate of DNA synthesis. In mammalian cells where there are long periods devoid of any DNA synthesis it may be necessary to have additional regulatory systems, affecting uptake for example, and thereby facilitating expansion of the precursor pools only when required.

Rubin: The deoxyribonucleotide pools vary enormously in animal cells according to their growth state; the ribonucleotide pools remain constant (Colby and Edlin 1970). The deoxynucleotide pools increase roughly in proportion to the rate of uptake of thymidine. As a result specific activity of the pool does not vary much, and [^{3}H]thymidine incorporation into DNA is a fairly good indicator of the rate of DNA synthesis in a culture. Obviously this is not true for [^{3}H]uridine and RNA.

Cunningham: As you point out, the ribonucleotide pools don't change much. We have shown (Cunningham and Pardee 1969) that the amount of uridine taken up by serum-stimulated confluent 3T3 cells is small compared to the total acid-soluble pool of uridine-containing compounds. Therefore, the type of "compensation" you mention for thymidine does not occur for uridine. In fact, we showed that increased uptake of [^{3}H]-uridine resulted in an equal increase in the specific activity of the uridine-containing compounds in the acid-soluble fraction.

Elsdale: Have you followed up the idea that the transport inhibitors are made only by confluent cells and not by non-confluent ones?

Cunningham: The inhibitor of phosphate transport is made only by confluent 3T3 cells, whereas the inhibitor of uridine transport is made by both non-confluent and confluent 3T3 and transformed 3T3 cells.

Elsdale: And in roughly the same amount?

Cunningham: Yes, it appears that non-confluent and confluent 3T3 and Py3T3 cells release about the same amount of inhibitor per mg of cell protein. This measurement was complicated by the fact that the cells not only release the inhibitor for uridine transport, but they also use up or inactivate the stimulating factor for uridine transport present in the serum. But the inhibitor is dialysable while the stimulating factor is not. Therefore, it was possible to estimate the amount of inhibitor released by assaying depleted medium both before and after dialysis.

Elsdale: The interesting thing from the point of view of growth regulation is why the phosphate transport inhibitor is made only by confluent cells.

Cunningham: Yes, but it might be a difference in release, not in synthesis of the inhibitor.

Clarke: Is there any evidence that in confluent cultures the uptake of phosphate is growth-limiting?

Cunningham: We have no evidence that the uptake of phosphate is growth-limiting. In fact, our experiments using [^{32}P]phosphate were done using medium in which the amount of phosphate was reduced 10-fold (to about 10^{-4} M) from the amount in Dulbecco's modified Eagle medium (about 10^{-3} M). This reduction in phosphate concentration has no effect on growth of the cells. We haven't tried lower concentrations, but I would imagine that it is present in excess in our experiments.

Rubin: I find it difficult to believe that phosphate is limiting in these cultures; it is present in buffering concentrations, not in substrate concentrations.

Dulbecco: I have to ask my traditional question about infection with mycoplasma! Your results resemble unpublished results that I once obtained with 3T3 cells that were attributable to uptake of thymidine by mycoplasma. I may point out that the effect of serum on thymidine transport in mycoplasma persists if the serum is partially digested with pronase, whereas the effect on cell growth disappears. This may be one way of distinguishing transport into cells from transport into mycoplasma. It would also be useful to look not just at total RNA but to see whether the 45S precursor is present, because if it is not present, you may be measuring mycoplasma RNA.

Also, I am worried by your experiments on the turnover of phospholipids, because you showed compartmentation.

Cunningham: Yes, there possibly could be a compartmentation effect. The answer to your question about mycoplasma infection is that it is very unlikely our cells were contaminated. They were checked in two other laboratories by the culture method and found to be negative. More importantly, we routinely checked the cells by the very sensitive [^{3}H]-thymidine radioautography method. Our cells were negative by this test. They showed grains only over the nuclei. Mycoplasma-infected cells show grains over the entire cytoplasm.

Dulbecco: It might be worthwhile to see if pronase-digested serum has any effect.

Todaro: A very sensitive test for mycoplasma seems to be adding labelled uridine (or thymidine) to cell cultures, taking the medium, precipitating with ammonium sulphate and then putting it on sucrose gradients. Almost all the mycoplasma strains band in isopycnic sucrose gradients at $1 \cdot 22$ to $1 \cdot 24$ g/cm^3. Using these methods, we have found a lot of cultures that have been considered to be negative, actually to be positive.

Cunningham: Were they negative by the [^{3}H]thymidine radioautography test?

Todaro: The relative sensitivity has not been directly compared.

Taylor-Papadimitriou: The radioautographic test is even more sensitive when the cells have been grown previously in low serum ($0 \cdot 5$ per cent) before being incubated with labelled thymidine.

Dulbecco: Mycoplasma appears to respond strongly to serum.

Todaro: By all the tests (including thymidine radioautography) the 3T3 cells in our laboratory have consistently been negative.

Rubin: Lewis Thomas (1969) said that every cell line that he examined contained mycoplasma.

Todaro: Every cell we obtain from certain laboratories has also been uniformly positive.

Cunningham: It is important when you do the radioautography test to label the cells very extensively with [^{3}H]thymidine so that the nuclei are quite black. We then consistently failed to find grains over the cytoplasm.

Stoker: We found that transformed cells seem to be infected with mycoplasma more frequently than untransformed cells of the same line; this is an additional reason for caution in any comparative studies.

REFERENCES

COLBY C. and EDLIN, G. (1970) *Biochemistry* **9**, 917–920.

CUNNINGHAM, D. D. and PARDEE, A. B. (1969) *Proc. natn. Acad. Sci. U.S.A.* **64**, 1049.

THOMAS, L. (1969) In *Cellular Recognition*, p. 139, ed. Smith, R. T. and Good, R. A. New York: Appleton-Century-Crofts.

TOBEY, R. A., ANDERSON, E. C. and PETERSEN, D. F. (1967) *J. Cell Biol.* **35**, 53.

WEBER, M. and RUBIN, H. (1971) *J. cell. Physiol.* in press.

ASPECTS OF MEMBRANE STRUCTURE AND FUNCTION IN *ESCHERICHIA COLI*

I. B. Holland, A. C. R. Samson, Éva M. Holland and B. W. Senior

Department of Genetics, University of Leicester

The bacterial plasma or cytoplasmic membrane is still a rather ill-defined and poorly understood structure. Nevertheless in recent years it has been increasingly realized that the membrane is a complex multi-functional organelle which is involved in a wide variety of essential cellular activities. Membrane functions include control of selective permeability and electron transport systems and oxidative phosphorylation. In this paper, however, we shall be concerned primarily with another functional aspect of the membrane, the metabolism of DNA. We shall also discuss the mechanism of action of colicins, protein antibiotics whose study is revealing some of the complexities of membrane structure and function. Finally, where possible the relevance of colicin studies to the role of the membrane in DNA metabolism in particular will be discussed.

STRUCTURE OF THE CELL ENVELOPE OF *ESCHERICHIA COLI*

The envelope of *Escherichia coli* is a complex multi-layered formation whose structure is generally accepted to be that shown in Fig. 1. Two unit membranes are visible in the electron microscope: the plasma membrane which encloses the cell cytoplasm and an outer membrane to which the various surface antigens are attached. The unit membranes contain lipid and protein in a configuration now thought to be rather different from that envisaged in the Danielli–Davson model (Danielli and Davson 1935). Instead of a continuous lipid bilayer coated with protein, considerable evidence suggests that the bilayer is interspersed with protein subunits which partially or fully penetrate the lipid. For simplicity we shall assume that the primary specificity of the membrane, in relation to the functional aspects to be discussed, resides in its protein content.

Although Fig. 1 depicts the *E. coli* surface as composed of discrete layers with the rigid layer in particular well separated from the plasma membrane, this is probably an oversimplification. Bayer (1968*a*, *b*) has shown that small ducts or extensions of the plasma membrane protrude into the rigid layer. These ducts, of which there are 200–400 per cell, become visible when bacteria are briefly plasmolysed, thus allowing the outer layers to separate from the plasma membrane. A strong correlation between the location of surface adsorption sites for bacteriophages T1 and T7 and the presence of the membrane tubules was found by Bayer, who suggests that these channels may reflect a common mechanism for the penetration of viral nucleic acids. In any event these results indicate that specific connexions between the rigid layer and the cytoplasmic membrane are present in the bacterial surface and that the plasma membrane has therefore at least some structural differentiation.

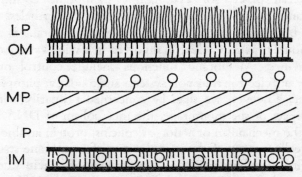

FIG. 1. Surface structure of *E. coli*. LP, lipopolysaccharide and surface antigens; MP, mucopeptide (rigid layer) covalently linked to trypsin-sensitive protein; P, periplasm containing many degradative enzymes; OM, outer lipoprotein membrane; IM, inner or plasma membrane. For further details see text and also Burge and Draper (1967) and Braun and Sieglin (1970).

THE PLASMA MEMBRANE AND DNA REPLICATION

Much of the evidence implicating the plasma membrane in several aspects of DNA metabolism has been reviewed previously (Ryter 1968; Holland 1970) and only the major findings will be noted here. It will be assumed that once DNA replication is initiated this process proceeds in a sequential manner by means of a replication fork, which traverses the bacterial chromosome from a specific starting point (the origin) to a specific terminus. Sueoka and Quinn (1968) have shown that the replication origin of the chromosome, which in bacteria can be defined genetically and biochemically, and the replication fork, are both physically bound to membrane-like structures. In addition Ryter (1968) has presented convincing electron-microscopic evidence for limited attachment sites between the plasma membrane and the bacterial chromosome. Finally several groups are now studying, in semi-*in vitro* systems, the mechanism of DNA synthesis in apparently membrane-bound systems (Smith, Schaller and Bonhoeffer 1970; Knippers and Strätling 1970). No definitive evidence for a specific membrane-DNA linkage has however yet been obtained.

MEMBRANE GROWTH AND SEGREGATION OF THE BACTERIAL CHROMOSOME

As indicated above, the bacterial chromosome is probably bound to the plasma membrane at at least two specific attachment sites. One site comprises the "replication apparatus" and probably reflects the direct participation of the membrane in the synthesis of DNA, perhaps analogous to the role of the ribosome surface in protein synthesis. In addition, Jacob, Brenner and Cuzin (1963) have postulated that specific attachment of the chromosome to the membrane is also essential to ensure precise segregation of daughter chromosomes and various plasmids at division. The suggested mechanism of segregation envisaged that growth of the bacterial membrane always commences in a central zone formed between the attachment sites of daughter chromosomes, and that as growth continues, the expanding membrane achieves separation of the chromosomes prior to a symmetrical division. Tentative support for this mode of growth of the plasma membrane (in *Bacillus subtilis*) was obtained by Jacob, Ryter and Cuzin (1966) who studied the pattern of distribution of cell membranes, labelled in the parental generation with potassium tellurite, amongst first and second generation progeny. Recently Donachie and Begg (1970) by microscopic observation of single cells of *E. coli* have demonstrated that a bacterial cell does grow in a manner exactly compatible with the mode of membrane growth postulated by Jacob and co-workers.

THE BACTERIAL GROWTH CYCLE

The initiation of chromosomal replication is a major event in the cell cycle and since this process and the associated process of cell division no doubt specifically involve the plasma membrane in some way, it is appropriate at this point to review briefly the main features of the bacterial growth cycle.

From the original work of Maaløe and Kjeldgaard (1966) and the more recent studies of Helmstetter and Cooper (1968; and see Helmstetter *et al.* 1968) it is now possible to visualize the chief stages of the bacterial growth cycle. These are illustrated in Fig. 2 for synchronously growing bacteria with a mean generation time of 45 minutes. Chromosomal replication is initiated around the middle of the cell cycle and is only completed in the daughter cells. Helmstetter and Cooper have also shown that over a wide range of growth rates the replication velocity, C,

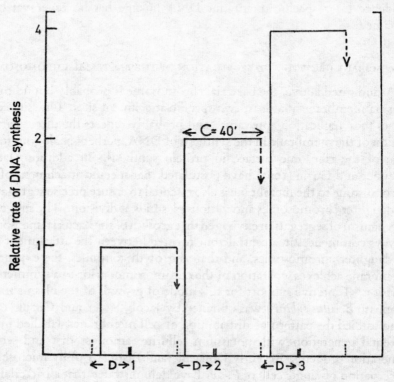

Fig. 2. Bacterial growth cycle. Idealized presentation of DNA replication and cell division in a bacterial culture growing synchronously in minimal-glucose medium with a generation time of 45 minutes at 37°C. The DNA replication velocity, C, is a constant and ensures that rounds of replication are completed in 40 minutes, independently of growth rate. The time (D) elapsing between termination of replication and division is also independent of growth rate, remaining constant at 20 minutes.

is constant at about 40 minutes. The initiation of chromosomal replication is therefore a strictly regulated event and the rate of DNA synthesis is governed only by the frequency of initiation and not directly by the bacterial growth rate. Although the cell age at which initiation of replication takes place varies with the growth rate, Donachie (1968) on the basis of Helmstetter and Cooper's results has calculated that the cell mass per chromosomal origin at which initiation occurs is constant.

Helmstetter and Cooper have defined two further parameters of bacterial growth (see Fig. 2). D is the time elapsing between completion of replication and division, which is also constant over a range of growth rates at about 20 minutes; the final parameter, I, is the time required for the synthesis or accumulation of some cell constituent of constant size per chromosomal origin, which promotes initiation. Since I is equal to the mean generation time and since C, the replication velocity, is constant, at fast growth rates new rounds of replication will be initiated before the completion of previous rounds. This will lead to the presence of multi-forked chromosomes, a situation which has been directly demonstrated by many workers. Conversely at slow growth rates where I is greater than C there will be periods in the growth cycle when DNA is not being synthesized.

We may now consider what factors are known to affect the initiation of replication and the subsequent development of the period D. Lark, Repko and Hoffman (1963) first showed that although replication cycles already in progress could be completed in the absence of further protein synthesis, the initiation of new rounds was dependent upon protein synthesis. In relation to D, Pierucci and Helmstetter (1969) have shown that completion of a round of replication is an essential requirement for division and in addition a *period* of protein synthesis equivalent to (and normally concurrent with) the period of DNA replication is also necessary for division. Entry into period D is therefore triggered by completion of a round of DNA replication and a period of *concurrent* protein synthesis. Finally, the fact that D is a constant suggests that synthetic activity associated with this period is, like replication, determined by the initial availability of some template and not by the availability of precursors.

Initiation of replication and cell division are therefore seen as controlled events, promoted by measuring devices which in one case trigger *single* new rounds of replication in response to each doubling of the cell mass or volume. Although the regulatory mechanisms, in particular those coordinating replication and the many components of the division process, are undoubtedly complex, they should be amenable to genetic analysis. Mutants in which replication and division are uncoupled (Inouye 1969), in which DNA is apparently synthesized "constitutively" (Yoshi-kawa and Haas 1968), or in which cell division is defective in a variety of ways (Hirota, Ryter and Jacob 1968) have already been isolated and should be helpful in elucidating the control mechanism.

MEMBRANE FUNCTION AND COLICINS

Unlike more conventional antibiotics colicins have extremely narrow antibacterial spectra. Moreover colicins are proteins with, in the case of

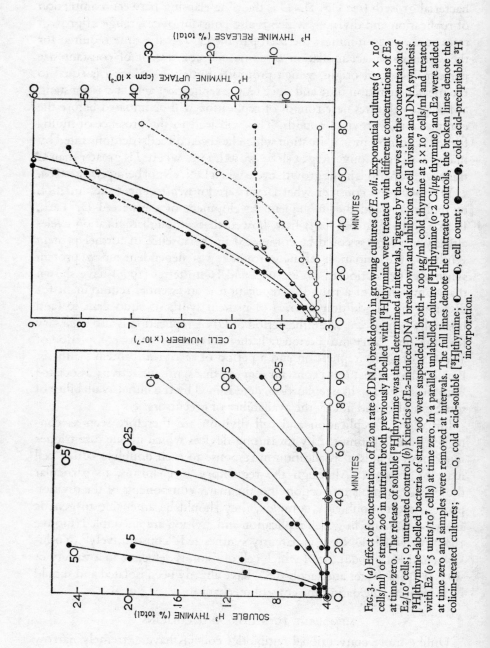

Fig. 3. (a) Effect of concentration of E2 on rate of DNA breakdown in growing cultures of *E. coli*. Exponential cultures (3×10^7 cells/ml) of strain 206 in nutrient broth previously labelled with [^{3}H]thymine were treated with different concentrations of E2 at time zero. The release of soluble [^{3}H]thymine was then determined at intervals. Figures by the curves are the concentration of E2/10^7 cells; o, untreated control. (b) Kinetics of E2-induced DNA breakdown and inhibition of cell division and DNA synthesis. [^{3}H]thymine-labelled bacteria of strain 206 were suspended in broth + 100 μg/ml cold thymine at 3×10^7 cells/ml and treated with E2 (o·5 units/10^7 cells) at time zero. In a parallel unlabelled culture [^{3}H]thymine (o·2 Ci/μg thymine) and E2 were added at time zero and samples were removed at intervals. The full lines denote the untreated controls, the broken lines denote the colicin-treated cultures; o——o, cold acid-soluble [^{3}H]thymine; ●——●, cell count; ●, cold acid-precipitable ^{3}H incorporation.

colicins E2 and E3, molecular weights of 60 000 (Herschman and Helinski 1967). The restricted range of activity of colicins is probably due to the requirement in sensitive bacteria for a specific surface receptor. This receptor can be readily lost or blocked by mutation and with the group E colicins a single mutational change of this kind can render a bacterium unable to adsorb, and therefore resistant to, colicins E1, E2, E3 and bacteriophage BF23 (Jenkin and Rowley 1955). The possible nature of the colicin E receptor will be considered later but first some of the primary consequences of colicin E action will be described.

Primary biochemical consequences of colicin fixation

Evidence obtained by F. Levinthal and C. Levinthal (cited by Luria 1964) indicated that colicin E1 induced the uncoupling of oxidative phosphorylation in sensitive bacteria. Macromolecular syntheses are consequently rapidly halted by treatment with E1. Later studies by Fields and Luria (1969a, b) confirmed these observations and also demonstrated that colicin E1 inhibited active transport, probably in consequence of the lowered ATP levels found in treated cells. Neither colicins K nor E1 have, however, been found to have any direct effects on membrane permeability. Similarly colicins E2 and E3 do not appear to induce any detectable changes in cell permeability (Nomura 1963; Luria 1964; Fields and Luria 1969a). Colicins E2 and E3 in fact act quite differently from E1 and the earliest detectable biochemical changes observed so far are degradation of DNA and the inhibition of protein synthesis respectively. As shown in Fig. 3a, the release of DNA breakdown products induced by E2 can be detected, depending upon the amount of E2 added, as early as 3 minutes after treatment. Although such experiments presumably measure the effects of a specific exonuclease activated by E2, recent work has also shown that endonucleolytic cleavage of DNA can be detected in the early stages of breakdown (Obinata and Mizuno 1970; Ringrose 1970a). It is not clear which of these nucleolytic activities occurs first in E2-treated cells but results (unpublished) indicate that at least two enzymes are involved and that both are specifically "activated" by E2 adsorption. RNA and protein synthesis are not immediately inhibited by E2 treatment and, as shown in Fig. 3b, DNA synthesis also continues for some time at the control rate although inhibition of cell division follows shortly after the onset of breakdown.

Colicin E3, which appears to attach to the same receptor as E2 (Maeda and Nomura 1966), is nevertheless without effect on DNA and RNA synthesis, but inhibition of protein synthesis commences 4–5 minutes after treatment and it is completely blocked after 11 minutes. This

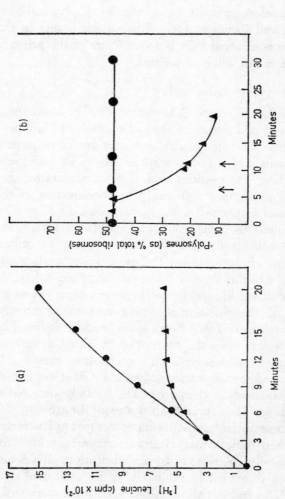

FIG. 4. (*a*) Effect of E3 on protein synthesis in strain A19. At zero time [³H]leucine (6 nCi/µg leucine) and 50 units/ml E3 were simultaneously added to exponential cultures (2 × 10⁸ cells/ml) of A19 growing in supplemented M9 medium at 37°C. Incorporation of radioactivity into acid-precipitable material was then determined at intervals. ●——●, control untreated culture; ▲——▲, E3-treated culture.

(*b*) Kinetics of the development of unstable (at 1 mм-Mg⁺⁺) polysomes in E3-treated cultures of A19. Exponential cultures (2 × 10⁸ cells/ml) of A19 growing in supplemented M9 medium at 37°C were treated with 50 units/ml E3 at zero time. Samples were removed at intervals from the E3-treated culture (▲——▲) and an untreated control culture (●——●) and the cells lysed. The lysates were centrifuged at 2°C on linear 15–30 per cent sucrose gradients prepared in TMA buffer (10⁻²м-tris, 5 × 10⁻²м-NH₄Cl; pH 7·2) containing 1мм-Mg⁺⁺ in a SW27 rotor for 2·5 hours at 27 000 rev/min. The gradients were analysed and the areas of polysome and ribosome peaks were measured; the percentage of polysomes in the total ribosomes was determined. Arrows mark the onset and completion of E3-directed inhibition of protein synthesis.

progressive inhibition of protein synthesis is precisely paralleled by the appearance of a physical change in cellular polysomes rendering them unusually fragile at low magnesium concentrations (Fig. 4). This change probably reflects an alteration in the 30S ribosomal component, since Konisky and Nomura (1967) showed that this subunit, when isolated from E3-treated cells, is defective in protein synthesis *in vitro*.

The evidence discussed so far indicates that colicins E1, E2 and E3 have profound effects on cellular functions located in the plasma membrane or in cell constituents closely associated with it. Evidence will be presented below which strongly suggests that colicins do not penetrate sensitive bacteria, but act extracellularly, at the plasma membrane surface.

Site of action of colicins

The interest in colicins as probes of membrane function stems from the observations by Nomura and co-workers (Nomura and Nakamura 1962; Nomura 1963) that the lethal action of colicin K could be reversed by treating affected cells with trypsin. Colicin K acts in a similar way to colicin E1 and Nomura was able to show that treatment with trypsin restored macromolecular synthesis and the capacity to form colonies in the inhibited cells. This demonstrated that colicin molecules remained at the cell surface and acted from the extracellular receptor, presumably through the intermediacy of the plasma membrane. Some support for this conclusion was obtained by Maeda and Nomura (1966) using radioactive E2. After adsorption of the colicin and subsequent breakage and fractionation of cellular components, more than 90 per cent of the labelled E2 was recovered from the envelope fraction and residual debris, while the soluble fraction contained at most 1 per cent of the E2. In a rather different approach Nomura and co-workers (Nomura 1964; Konisky and Nomura 1967) and Ringrose (1970a) have shown that E2 and E3 have no measurable effects *in vitro* upon *E. coli* DNA and ribosomes respectively. The integrity of the whole cell therefore appears necessary for the colicin to act.

In contrast to the action of colicin K, rescue of viable cells by trypsin after treatment with E2 and E3 is normally only possible for a few minutes after adsorption of colicin. Nomura (1967) has suggested that this probably reflects the irreversible biochemical changes promoted by these colicins at the DNA or ribosomal level, rather than a failure to reverse changes in the membrane, as found possible with colicin K. In support of this, Fig. 5 shows that early addition of trypsin to cultures treated with E2 and in which exonucleolytic breakdown has already commenced, immediately inhibits further breakdown. Various control experiments have clearly demonstrated that this is due to the proteolytic

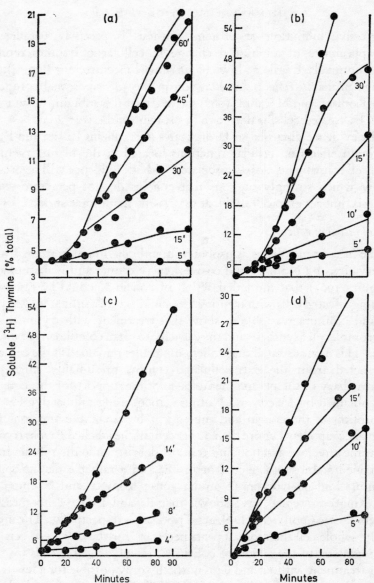

Fig. 5. Effect of trypsin on DNA breakdown induced by E2. Cultures of strain 206 prelabelled with [³H]thymine were washed and resuspended in nutrient broth pH 7·8 (*a* and *b*) or in tris phosphate buffer pH 7·8 (*c* and *d*) containing 10^{-2}M-KH_2PO_4, 10^{-3}M-$MgSO_4$ and 10^{-4}M-$CaCl_2$. Primary cultures treated with E2 throughout are designated as controls. Samples were removed from the primary cultures and treated with 1 mg/ml trypsin at the times shown on the curves. (*a*) Nutrient broth culture treated with E2 (2 units/10^7 bacteria) at zero time and treated with trypsin at the times indicated. (*b*) Nutrient broth cultures treated with E2 (0·2 units/10^7 bacteria). (*c*) Cells suspended in tris phosphate buffer and treated with E2 (8 units/10^7 bacteria). (*d*) Pre-adsorption of E2 (10 units/10^7 cells) for 45 minutes at 4°C; cultures then centrifuged and resuspended in tris phosphate buffer before trypsin treatment.

digestion of the adsorbed colicin and not to any non-specific effects. The inhibitory effect of trypsin is observed with both growing and resting cells and is independent of whether colicin is pre-adsorbed (at 4°C) or not. These results strongly suggest that removal of colicin by trypsin halts nucleolytic activity already initiated and does not merely prevent the formation of further active surface complexes. The progressive increase in resistance of the breakdown process to trypsin is not yet understood but may reflect an increasing non-specificity of the degradative enzymes. The above interpretation of the mechanism of trypsin inhibition has been confirmed recently by Ringrose (1970*b*). He has shown that the continued formation of single-strand breaks in DNA can be halted, and the breaks repaired, when trypsin is added to cells shortly after treatment with E2. It seems reasonable to conclude, therefore, that the activity of a membrane-bound nuclease(s) is promoted by adsorption of E2 to the cell surface, but is suppressed again when E2 is removed by trypsin.

Nature of the colicin receptor

Considerable genetic evidence indicates that colicins adsorb to bacteria through the agency of specific receptors analogous to bacteriophage receptors (Fredericq 1957). Adsorption of colicin is very rapid and is normally irreversible even at 4°C. Furthermore, resistant mutants which fail to adsorb specific colicins can be readily isolated. Extracts from such resistant bacteria do not neutralize colicin whereas extracts from wild-type sensitive strains do (Bordet and Beumer 1948, 1951). In addition, antibacterial sera, although without direct effect upon colicin, protect sensitive cells against their action (Bordet 1948). In a quantitative study of the adsorption of E2 using radioactively labelled colicin, Maeda and Nomura (1966) calculated that a sensitive cell adsorbs a maximum of 2000–3000 colicin molecules, although a lethal hit (see below) required the adsorption of about 100 colicin molecules per cell.

Although the available evidence suggests that colicin receptors are specific components of the cell surface, no attempt has yet been made to isolate and characterize the receptor or to determine its precise location in the cell surface. Some results from our own studies (unpublished) indicate that the receptors for colicins E2 and E3 are resistant to trypsin and lie in the murein or rigid layer of the cell envelope. Studies by other workers, however (Šmarda and Taubeneck 1968; Bhattacharyya *et al.* 1970), suggest that the colicin receptor may be part of the plasma membrane itself. Purification and characterization of the receptor are now needed to resolve this problem.

Single-hit killing by colicins

Jacob, Siminovitch and Wollman (1952) first showed that the killing action of colicin E1 was a single-hit process analogous to phage killing. This has been confirmed for a variety of colicins including E2 (Shannon and Hedges 1967). These findings indicate that the adsorption of a single colicin molecule can cause death of a bacterial cell, although, as already noted above, the probability per adsorbed molecule of producing a lethal hit may be as low as one in a hundred. The efficiency or probability of a lethal hit is strongly influenced by the physiological state of the cells (unpublished results). However, the basis of this feature of colicin action is not understood, although receptor heterogeneity, low affinity of receptor-bound colicin for sensitive membrane sites, or simply a low probability of colicin–membrane interactions actually promoting a stable alternative state of the plasma membrane, may all be advanced as possible explanations.

Mutants refractory to colicin

In order to explain the single-hit killing and extracellular action of colicins, Nomura (1964) proposed the following model for colicin action. Fixation of colicin to the surface receptor initiates a specific stimulus which is propagated along the membrane by a specific transmission system, finally producing a characteristic biochemical change at a sensitive site (or target). A major prediction of the model is that specific steps in the action of colicin, subsequent to adsorption, should be demonstrable by genetical or biochemical means. In fact several genetically distinct classes of mutant which still adsorb colicin normally, but are refractory (insensitive) to its effects, have been isolated (Hill and Holland 1967; Nagel de Zwaig and Luria 1967; Nomura and Witten 1967).

The properties of some of these mutants will now be discussed. Mutants refractory to colicin E are of two major types, those specifically refractory to a single colicin, for example to E2 or E1, and those which are refractory to two or more E colicins and often to other, unrelated colicins also. In contrast to mutations which lead to loss of colicin binding, mutations to refractivity in almost all cases studied so far are pleiotropic. Refractory mutants display a variety of altered surface properties although in the one case examined no change in surface lipopolysaccharide could be detected (Nagel de Zwaig and Luria 1967). The properties of all these mutants suggest therefore that they have altered membranes.

Among multirefractory mutants Nomura and Witten (1967) have described a strain, refractory to E2 and E3 at 42°C, which becomes immediately sensitive on shifting to low temperature. It was suggested therefore that this mutant may produce a temperature-sensitive protein directly concerned in some post-adsorption step in the action of colicin. Nagel de Zwaig and Luria (1969) have described a temperature-sensitive mutant (tol VIII) which is refractory to all E colicins and which behaves in a rather different way to the Nomura strain. In these mutants changes in colicin sensitivity, after temperature shift, depend in a rather complex way on bacterial growth. Thus when colicin is added to the bacteria before temperature shift, the phenotype prevailing before the shift is maintained. Moreover when untreated cells undergo a shift-down from 41° to 30°C, the kinetics of the appearance of sensitivity to colicin added subsequently indicate that only adsorption to receptors synthesized after the shift could promote lethal effects. Thus the colicin receptor and the surface component controlled in some way by the *tol* gene apparently segregate together during growth. A curious feature of these mutants was observed, however, in temperature shift-up experiments. In this case the kinetics of the disappearance of receptors capable of promoting the lethal effect of freshly added E1 were too fast to be accounted for by cell growth alone. Under these conditions, therefore, the picture may be complicated by rapid turnover of the surface element controlled by the *tol* gene. Although Nagel de Zwaig and Luria were able to conclude that the gene product of the *tol* VIII gene was a protein, it was not possible to identify this protein either as a component of the membrane or as an enzyme involved in membrane (or envelope) biosynthesis.

Mutants specifically refractory to colicin E2 (Cet) have also been extensively studied (Holland 1967, 1968). These strains are almost all cold-sensitive mutants, since they are refractory at low temperature but largely sensitive at high temperature. One mutant of this type (CetB) has been studied in temperature shift experiments and found to undergo an immediate change in phenotype independently of the direction of the shift. This indicated that the product of the *cet*B gene is a protein directly involved in the action of E2 which in the mutants undergoes a reversible conformational change at low temperature, thereby blocking E2 action. Alternatively these mutants form an altered membrane (at both 30° and 40°C) which is unable to promote the action of E2 at low temperature but is capable of a lethal response to adsorbed colicin at 42°C. The analysis of envelope proteins in these mutants (described below) supports this latter alternative.

ANALYSIS OF ENVELOPE PROTEINS OF *ESCHERICHIA COLI*

In contrast to investigations of the protein content of animal cell membranes, the bacterial plasma membrane has received little attention so far.

Such studies, which are essential to any understanding of the functional role of the membrane, are, however, now being initiated in several laboratories. Envelope or membrane fractions may be prepared by physical breakage of cells, followed by differential high-speed centrifugation and thorough washing in buffer of low ionic strength. The membrane material obtained, constituting 4–6 per cent of the total cell protein and essentially free of RNA and soluble proteins, is then dispersed in 0·1 per cent sodium dodecyl sulphate and analysed on polyacrylamide gel. As shown in Fig. 6a, when the total envelope protein of *E. coli* is analysed in this way about 13 major polypeptide peaks are obtained, although some peaks are clearly complex. Better resolution is obtained if lipid is removed before electrophoresis and when the protein is radioactively labelled. In these conditions at least 20–30 peaks may be detected (Schnaitman 1970). Nevertheless the profiles shown in Fig. 6 are quite reproducible and are obtained whether the bacteria are grown in complex or simple media and at either high or low temperatures. A major feature of the gels is the large peak *g*, a polypeptide with a molecular weight of about 40 000 and constituting about 30 per cent of the total envelope protein. Similar results have been obtained by Schnaitman (1970) who has also shown, by density-gradient fractionation of the envelope material before analysis, that this predominating polypeptide is probably associated with the outer membrane, of which it constitutes 70 per cent of the total protein.

These techniques of acrylamide gel analysis have been applied to various Cet mutants and mutants resistant to colicin through loss of the surface receptor. With the latter, no differences between wild-type and mutant strains have been detected. As shown in Fig. 6a(2), however, analysis of CetB mutants demonstrates that the content of a specific peak *e* shows a two-fold increase in the mutant compared to the parent strain. Identical results were obtained whether the mutants were grown at either 25° or 42°C before analysis and therefore refractivity to E2, which is maximally expressed at low temperature, is not directly correlated with increased amounts of polypeptide *e* in the cell surface. Rather it seems that the altered cell membrane, which causes or which results from excess incorporation of polypeptide *e*, is unable to respond to E2 at low temperature but is able to transmit the lethal effects of the colicin at high temperature. The mobility of polypeptide *e* indicates a molecular weight of about 43 000 daltons and calculation shows this polypeptide to constitute about 3 per cent of the total envelope protein. Two refractory mutants carrying a *cet*C locus have also been studied and in both cases the same two-fold increase in peak *e* is observed (Fig. 6). These mutants are also maximally refractory to E2 at 25°C but in this case the mutation to refractivity is pleiotropic. In addition to having altered surface properties

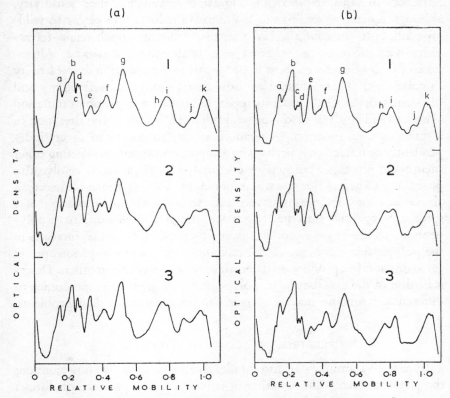

FIG. 6. Analysis of cell envelope proteins. Envelope fraction obtained from sonicated bacteria, purified by differential centrifugation and thorough washing with 10⁻² M-sodium phosphate buffer. Final pellets, free of RNA and soluble protein, were dissolved in 1 per cent sodium dodecyl sulphate in the presence of 1 per cent 2-mercaptoethanol and dialysed against 0·2 per cent sodium dodecyl sulphate (0·1 per cent 2-mercaptoethanol). 100 μg of envelope protein were applied to 10 per cent acrylamide, 0·1 per cent sodium dodecyl sulphate gels at pH 7·0 and electrophoresed. Gels were stained with coomassie blue and scanned with a Joyce-Loebl microdensitometer. Strain characteristics: (*a*) 1 = ASH1 wild-type *cet*⁺; 2 = ASH101 *cet*B; 3 = ASH111 *cet*C; (*b*) 1 = ASH114 *cet*C; 2 = ASH706, a UV-resistant revertant of ASH114 which is still refractory to E2; 3 = ASH711, an E2-sensitive revertant of ASH114 which is still sensitive to UV. Letters denote major reproducible peaks with molecular weights decreasing to the right.

(e.g. sensitivity to detergent), CetC mutants are sensitive to UV, recombination deficient and filament forming or resistant to bacteriophage λ (Holland *et al.* 1970).

Various revertants of the CetC mutant, ASH114 have been studied and have confirmed the specificity of the observed envelope changes in refractory mutants. As shown in Fig. 6*b*, revertants to colicin sensitivity, although remaining sensitive to UV, show a reduction in peak *e* to wild-type amounts. In contrast, UV-resistant revertants which retain refractivity to E2 also retain the enlarged peak *e* in all strains tested so far. Alterations of peak *e* therefore appear to be specific to the mutation to refractivity to colicin E2. These results also show that although E2-refractivity and UV-sensitivity in CetC mutants appear to arise by a single point mutation (Threlfall and Holland 1970), the sensitivity to UV reflects the alteration of a quite separate function. Alterations in this function can revert independently of refractivity, perhaps by a suppressor mutation affecting other membrane proteins. Further genetic analysis is required to resolve this problem. Other major questions posed by the experiments described above are: the mechanism which leads to increased incorporation of a specific polypeptide into the cell surface; the precise relationship of polypeptide *e* to the *cet* gene product; and why apparently similar increases in the polypeptide *e* content of cell envelopes are variously pleiotropic in some mutants but produce no detectable secondary effects in others. Determination of the distribution of polypeptide *e* in the plasma membrane of different Cet mutants may be necessary to resolve some of these problems.

MECHANISM OF ACTION OF COLICIN E2

By way of a summary of current ideas on colicins we shall now consider the possible mechanism of action of colicin E2. The presence of distinct steps subsequent to adsorption, as predicted by the Nomura model, has been confirmed. For simplicity however we prefer to regard this process as a "biochemical" pathway with, it is hoped, identifiable intermediates. Thus in the first instance colicin E2 is presumed to penetrate the outermost lipoprotein layer of the bacterial surface and to bind irreversibly to specific receptors, located either in the rigid layer or on the surface of the cytoplasmic membrane. The formation of this initial complex (I) can take place at 4°C in tris phosphate buffer. The formation of complex I is not lethal, however, and it does not lead to detectable breakdown of DNA. Complex II is a specific colicin–membrane complex, formed by a minority of the bound E2 molecules, which promotes rapid DNA breakdown. The efficiency and timing of the formation of complex II depends upon temperature, the concentration of E2 (cf. Fig. 3), the rate of bacterial growth, extracellular KH_2PO_4 and the presence of specific membrane (or envelope) proteins (see also Holland and Holland 1971).

Protein and DNA synthesis are not required for the formation of complex II. Recent work (P. Reeves, personal communication) in fact suggests that at least in some conditions complex II is formed in the absence of energy metabolism. The continued functioning of an E2-specific exonuclease does, however, require an available energy source (unpublished results). Complex II is envisaged as an altered conformational state of a particular part of the bacterial membrane, induced locally by E2 and perhaps propagated by secondary conformational changes of membrane proteins. In consequence of this molecular rearrangement in the cell surface, specific membrane-bound DNases, previously quiescent, now attack the DNA. Digestion of the colicin by trypsin destroys complex II and the membrane can then return to the normal state and the primary nucleolytic activity is halted.

The mode of action of E2 and the properties of mutants specifically refractory to colicin suggest that the particular part of the plasma membrane affected by the colicin is normally involved in DNA metabolism. An attractive candidate for the effective site of E2 action is therefore that part of the membrane which binds the chromosome and which forms part of the replication apparatus. Fuchs and Hanawalt (1970) have recently described a procedure for the isolation of a membrane–DNA fraction with properties expected of such a replication complex. Comparison of the protein constituents of this membrane fraction with those of the total plasma membrane in both wild-type and Cet mutants should provide a crucial test of our hypothesis. It is hoped that such studies would in addition reveal the presence of any unique proteins in the membrane component which makes up the DNA binding site(s).

In regard to the normal cellular function of the DNases activated by colicin E2, *E. coli* mutants each defective in one of several nucleases have been screened but this has revealed no correlation with the action of E2. In particular, despite the fact that CetC mutants are both UV-sensitive and recombination deficient, mutants specifically defective in UV-repair or recombination are fully sensitive to E2. The possibility remains however that adsorption of E2 activates a battery of membrane-bound repair or replication nucleases, any one of which can promote massive breakdown of DNA and cell death.

CONCLUSIONS

The picture now emerging of the bacterial plasma membrane is that of a complex organelle differentiated in some way into a number of functional parts. Despite this complexity, preliminary analyses indicate that more than 90 per cent of the membrane protein is made up of only 20–30 distinct polypeptides (or groups of polypeptides of similar size). Conceivably these "structural" proteins, either uniformly or discretely distributed, form a framework or background upon which the great multiplicity of membrane-associated enzymes is organized (see also Luria 1969). Prospects for the functional analysis of some of these major membrane proteins appear to be good, since they are altered or lost in Cet mutants, in multi-refractory mutants (Schnaitman, personal communication), in filamentous forms of *E. coli* B (Weinbaum, Fischman and Okuda 1970) and in mutants of *E. coli* (defective in DNA replication) in conditions where DNA synthesis is blocked (Inouye and Guthrie 1969; Shapiro *et al.* 1970). In contrast Schnaitman (1970) has shown that shifting *E. coli* from aerobic to anaerobic conditions, although having a marked effect upon enzyme activity associated with the membrane, does not affect the pattern of "structural proteins".

The study of the mode of action of colicins continues to aid the understanding of the functional organization of the plasma membrane. In particular, colicins reveal something of the specificity of protein interactions in membranes and the mechanism of growth of membrane and other surface layers. Colicin studies also suggest that some intracellular enzymes, for example, some concerned in DNA metabolism and perhaps cell division, are membrane bound; that the distribution of these enzymes within the membrane contributes to their specificity; and finally that specific alterations in membrane structure can activate or deactivate such enzymes. Regulatory mechanisms of this kind may be of paramount importance in the control of biosynthetic activity at the membrane surface, particularly in relation to discontinuous events like division and the initiation of DNA replication. Thus specific conformational changes and conceivably differential release and binding of regulatory proteins, within a linearly expanding cell membrane, may facilitate the triggering of chromosomal replication and division at precise intervals.

SUMMARY

The DNA in bacteria is attached to the cell membrane and replication proceeds sequentially as the chromosome passes through a membrane-bound replication apparatus. Initiation of replication is a precisely controlled event and is probably triggered by some device which responds to the attainment of a specific cell mass, which is constant for a given growth rate. Membrane growth appears to proceed linearly from a central zone and accomplishes segregation of daughter chromosomes. It is suggested that the physico-chemical changes in the expanding cell membrane may also play a role in determining the timing of discontinuous events like division and the initiation of chromosomal replication.

The use of colicins as probes of membrane function is described. Such studies reveal some of the complex interactions between "structural" proteins in the membrane and how the activities of some membrane-bound enzymes may be regulated. Also in conjunction with colicin studies, envelope proteins in *E. coli* have been analysed. A relatively small number of major polypeptides are present, one of which is specifically altered as a result of a mutation to refractivity to colicin E2.

Acknowledgements

Part of this work was supported by a grant (G.967/194/B) from the Medical Research Council; B.W.S. gratefully acknowledges receipt of a Beit Memorial Fellowship for Medical Research; A.C.R.S. gratefully acknowledges receipt of a Science Research Council Studentship. We are also extremely grateful to Dr Carl A. Schnaitman for pre-prints of his papers on *E. coli* membranes and to Professor R. H. Pritchard for many invaluable discussions.

REFERENCES

BAYER, M. E. (1968*a*) *J. Virol.* **2**, 346.
BAYER, M. E. (1968*b*) *J. gen. Microbiol.* **53**, 395.
BHATTACHARYYA, P., WENDT, L., WHITNEY, E. and SILVER, S. (1970) *Science* **168**, 998.
BORDET, P. (1948) *C. r. Séanc. Soc. Biol.* **142**, 257.
BORDET, P. and BEUMER, J. (1948) *C. r. hebd. Séanc. Acad. Sci., Paris, Sér. D* **142**, 259.
BORDET, P. and BEUMER, J. (1951) *Revue belge Path. Méd. exp.* **21**, 245.
BRAUN, V. and SIEGLIN, U. (1970) *Eur. J. Biochem.* **13**, 336.
BURGE, R. E. and DRAPER, J. C. (1967) *J. molec. Biol.* **28**, 173.
DANIELLI, J. F. and DAVSON, H. (1935) *J. cell. comp. Physiol.* **5**, 495.
DONACHIE, W. D. (1968) *Nature, Lond.* **219**, 1077.
DONACHIE, W. D. and BEGG, K. J. (1970) *Nature, Lond.* **227**, 1220.
FIELDS, K. L. and LURIA, S. E. (1969*a*) *J. Bact.* **97**, 57.
FIELDS, K. L. and LURIA, S. E. (1969*b*) *J. Bact.* **97**, 64.
FREDERICQ, P. (1957) *A. Rev. Microbiol.* **11**, 7.
FUCHS, E. and HANAWALT, P. (1970) *J. molec. Biol.* **52**, 301.
HELMSTETTER, C. E. and COOPER, S. (1968) *J. molec. Biol.* **31**, 507.
HELMSTETTER, C. E., COOPER, S., PIERUCCI, O. and REVELAS, E. (1968) *Cold Spring Harbor Symp. quant. Biol.* **33**, 809.
HERSCHMAN, H. R. and HELINSKI, D. R. (1967) *J. biol. Chem.* **242**, 5360.
HILL, C. and HOLLAND, I. B. (1967) *J. Bact.* **94**, 677.
HIROTA, Y., RYTER, A. and JACOB, F. (1968) *Cold Spring Harbor Symp. quant. Biol.* **33**, 677.
HOLLAND, É. M. and HOLLAND, I. B. (1971) *J. gen. Microbiol.* in press.
HOLLAND, I. B. (1967) *Molec. gen. Genetics* **100**, 242.
HOLLAND, I. B. (1968) *J. molec. Biol.* **31**, 267.
HOLLAND, I. B. (1970) *Science Prog.* **58**, 71.
HOLLAND, I. B., THRELFALL, E. J., HOLLAND, É. M., DARBY, V. and SAMSON, A. C. R. (1970) *J. gen. Microbiol.* **62**, 371.
INOUYE, M. (1969) *J. Bact.* **99**, 842.
INOUYE, M. and GUTHRIE, J. P. (1969) *Proc. natn. Acad. Sci. U.S.A.* **64**, 957.
JACOB, F., BRENNER, S. and CUZIN, F. (1963) *Cold Spring Harbor Symp. quant. Biol.* **28**, 329.

JACOB, F., RYTER, A. and CUZIN, F. (1966) *Proc. R. Soc. B* **164**, 267.

JACOB, F., SIMINOVITCH, L. and WOLLMAN, E. L. (1952) *Ann. Inst. Pasteur* **83**, 295.

JENKIN, C. R. and ROWLEY, D. (1955) *Nature, Lond.* **175**, 779.

KNIPPERS, R., and STRÄTLING, W. (1970) *Nature, Lond.* **226**, 713.

KONISKY, J. and NOMURA, M. (1967) *J. molec. Biol.* **26**, 181.

LARK, K. G., REPKO, T. and HOFFMAN, E. J. (1963) *Biochim. biophys. Acta* **76**, 9.

LURIA, S. E. (1964) *Ann. Inst. Pasteur*, suppl. Nov. **107**, 67.

LURIA, S. E. (1969) *Ciba Fdn Symp. Bacterial Episomes and Plasmids*, pp. 258–260. London: Churchill.

MAALØE, O. and KJELDGAARD, N. O. (1966) *Control of Macromolecular Synthesis*. New York: Benjamin.

MAEDA, A. and NOMURA, M. (1966) *J. Bact.* **91**, 685.

NAGEL DE ZWAIG, R. and LURIA, S. E. (1967) *J. Bact.* **94**, 1112.

NAGEL DE ZWAIG, R. and LURIA, S. E. (1969) *J. Bact.* **99**, 78.

NOMURA, M. (1963) *Cold Spring Harbor Symp. quant. Biol.* **28**, 315.

NOMURA, M. (1964) *Proc. natn. Acad. Sci. U.S.A.* **52**, 1514.

NOMURA, M. (1967) *A. Rev. Microbiol.* **21**, 257.

NOMURA, M. and NAKAMURA, M. (1962) *Biochim. biophys. Acta* **7**, 306.

NOMURA, M. and WITTEN, C. (1967) *J. Bact.* **94**, 1111.

OBINATA, M. and MIZUNO, D. (1970) *Biochim. biophys. Acta* **199**, 330.

PIERUCCI, O. and HELMSTETTER, C. E. (1969) *Fedn Proc. Fedn Am. Socs exp. Biol.* **28**, 1755.

RINGROSE, P. (1970a) PhD. Thesis, University of Cambridge.

RINGROSE, P. (1970b) *Biochim. biophys. Acta* **213**, 320.

RYTER, A. (1968) *Bact. Rev.* **32**, 39.

SCHNAITMAN, C. A. (1970) *J. Bact.* **104**, 882.

SHANNON, R. and HEDGES, A. J. (1967) *J. Bact.* **93**, 1353.

SHAPIRO, B. M., SICCARDI, A. G., HIROTA, Y. and JACOB, F. (1970) *J. molec. Biol.* **52**, 75.

ŠMARDA, J. and TAUBENECK, U. (1968) *J. gen. Microbiol.* **52**, 161.

SMITH, D. W., SCHALLER, H. E. and BONHOEFFER, E. J. (1970) *Nature, Lond.* **226**, 711.

SUEOKA, N. and QUINN, W. G. (1968) *Cold Spring Harbor Symp. quant. Biol.* **33**, 695.

THRELFALL, E. J. and HOLLAND, I. B. (1970) *J. gen. Microbiol.* **62**, 383.

WEINBAUM, G., FISCHMAN, D. A. and OKUDA, S. (1970) *J. Cell Biol.* **45**, 493.

YOSHIKAWA, H. and HAAS, M. (1968) *Cold Spring Harbor Symp. quant. Biol.* **33**, 843.

DISCUSSION

Elsdale: Am I right in thinking that colicin does not itself interact *in vitro* with DNA? Secondly, I take it that your resistant *E. coli* mutants are single cistron changes? Following on this is the more general question of whether you see any analogy between the colicin system and the interferon system in animal cells.

Holland: We can demonstrate no effects of colicin E3 on protein synthesis *in vitro*. We know that E3 induces a specific change in the ribosome *in vivo* but it doesn't interact with ribosomes *in vitro*. E2 doesn't seem to have any *in vitro* interaction with DNA.

All the mutants studied, including the very complicated pleiotropic ones, appear to arise from single point mutations.

Some reports have previously suggested that interferon acts at the cell surface (Friedman 1967), and Levy (see Levy, Baron and Buckler 1969) has reported that ribosomes from interferon-treated cells discriminate between viral and host cell RNA *in vitro*. If these reports are confirmed this would indicate a clear analogy between colicin action and that of interferon.

Taylor-Papadimitriou: I don't think there is any good evidence that interferon acts at the cell surface, and I think the effect on ribosomes is still rather obscure.

Burger: Is the mutant protein really in the inner membrane or is it in the outer wall layers? You could approach the question by studying whether this protein is still found in sphaeroplasts or whether it can be found in the Cota-Robles material which is wall-like and is released into the incubation medium (Birdsell and Cota-Robles 1967).

Holland: In our analysis of surface proteins of *E. coli* we have so far only examined the whole envelope. As already indicated, Schnaitman (1970) has described a method for the partial fractionation of the surface layers of *E. coli* and using this we hope to locate precisely the position of polypeptide *e*. Our analytical procedures will in fact only detect proteins which are tightly bound to the cell surface but we cannot exclude the possibility that polypeptide *e* is also bound to structures which can be secreted into the medium.

Warren: Could your electrophoretic pattern be changed by the addition of colicins to the proteins before electrophoresis?

Holland: We presume that colicin is present in the envelope fraction from treated bacteria. However, the maximum amount of E2 protein calculated to be present would not be sufficient to produce an observable peak on the gels by our present methods.

Warren: You would think that it might combine with some of the proteins rather specifically. You could label the colicin radioactively and see if binding occurs.

Holland: This is a possibility.

Clarke: Is there any evidence that the test cell itself produces a colicin, and has it any function in the cell?

Holland: Strains of *E. coli* (*col* E2$^+$) which produce colicin often do have colicin in the bacterial cell surface. The functional significance of this is obscure, however, but it is not, for example, essential for the expression of immunity of the producing strain to its own colicin. In most cases, including colicin E2 and E3, the nature and location of the binding site in the surface of the producing bacteria is unknown.

Pitts: Is this a case of competition between the colicin and a normal cellular protein for a site, or does the colicin just add to the completed membrane?

Holland: It seems quite likely that E2 can successfully interact with a completed membrane. The colicin is then either incorporated or acts at the membrane surface in some way to promote a specific conformational change in one or more membrane proteins.

Pitts: Can you competitively inhibit the adsorption of colicin to the cell with normal membrane components?

Holland: We haven't tried that. I should emphasize, however, that although we believe that E2 promotes specific changes in the plasma membrane we have no evidence that the colicin-binding site and the sensitive site in the cell membrane are necessarily the same. Experiments designed along the lines you suggest would therefore have to discriminate between competitive inhibition of either adsorption of E2 or the lethal action of colicin, subsequent to adsorption.

Dulbecco: You said that there is some change in the *E. coli* membrane when DNA synthesis is stopped. What is the change?

Holland: This evidence comes from work done by Hirota and his colleagues (Y. Hirota, personal communication) who have found that when DNA synthesis in *E. coli* is inhibited, at least one polypeptide is displaced from the surface membranes, as revealed by gel electrophoresis of envelope preparations.

Burger: If there are pleiotropic mutants which are UV-sensitive, and at the same time pleiotropic mutants which affect the formation of filaments (i.e. septum formation), has anyone looked for colicins which block the formation of septa?

Holland: Only colicin E2 has so far been reported to block cell division. The specificity of this effect is not yet clear but it is apparently not due to the inhibition of DNA synthesis, which is only observed several minutes after cell division is affected. We cannot exclude the possibility, therefore, that E2 has two independent effects on sensitive bacteria—the induction of DNA breakdown and the inhibition of cell division.

Burger: Usually people stress the fact that regulation of cell division may be based on cell mass; I think that one should consider at the same time the relation between mass and surface area and that it may be more the surface–mass relation which is important in triggering cell division.

Holland: Yes, I agree. I think the best models, for example that worked out by R. H. Pritchard, are based on either a volume doubling or surface area doubling, but cell mass is often the simplest thing to measure.

Dulbecco: It seems to me that everything you said about the connexion between DNA synthesis and cell replication is also compatible with the notion that division does not depend directly on the completion of a round of replication, but on the initiation of the next round. Is there a reason for separating the two?

Holland: In a sense the initiation of DNA replication is the event which puts in motion the whole bacterial growth cycle, including the division sequence. Nevertheless there is abundant evidence, as discussed earlier, that completion of an undisturbed round of DNA replication is also required for subsequent cell division.

Dulbecco: I prefer the second idea, that initiation is the key factor, because there is then generality of the relation of protein synthetic periods to DNA replication as in phage, and in all animal viruses that contain DNA. There the completion of DNA synthesis does not seem to be required for initiating "late" protein synthesis. The essential thing seems to be that the replication fork reaches some particular gene whose function is crucial. This suggests that the semi-denatured DNA, at the replication fork, can be preferentially transcribed, generating the signal for late protein synthesis. So initiation of the next round of DNA synthesis may be a better candidate for replication.

Holland: Recent work by Pierucci and Helmstetter (1969) has now shown that DNA replication must be completed concomitantly with an equivalent period of protein synthesis. If these two processes are uncoupled division is delayed.

Dulbecco: But when you do this you also initiate the next round of DNA synthesis, so one cannot separate the two.

This question of a relationship to mass is a very complicated hypothesis; it seems very imprecise. It is more likely that the regulation is carried out by the synthesis of one specific protein like sigma, which then activates other processes.

Holland: The complete execution of the control system may be complicated but the relationship—that as the cell mass doubles replication is initiated—seems quite simple.

Dulbecco: But that is not simple, because the doubling of mass is not a simple thing. The simple situation is having one gene activated and one protein made, but doubling the mass doubles everything. So it is not simple at all.

Holland: The operation of the control system, once the triggering event has occurred, certainly does not seem to be simple. When repression of initiation is lifted, one does not see continuous re-replication of the chromosome but only the execution of a single round. Additional factors in the regulatory mechanism are necessary to ensure that re-initiation does not occur before completion of the round.

Dulbecco: But you can initiate before the round is complete.

Holland: Yes, but only when the cell mass has doubled.

Stoker: Doubling in mass may always be a reflexion of an out-of-step doubling of something else; there may be one protein which is diluted out, and whose probability of interaction with a target is altered when the mass is doubled.

Dulbecco: I don't find the evidence compelling.

Clarke: Is there any evidence to support the model proposed for the control of DNA synthesis in bacteria by Rosenberg, Cavalieri and Ungers (1969)? This suggests that the initiation of DNA synthesis is accompanied by the production of a unit of inhibitor, which prevents reinitiation. This inhibitor is inactivated by a protein made in such a manner that one unit is produced only when the cell protein has doubled.

Holland: This is still speculation. The available theories are that on the one hand, while the cell mass doubles a new replication fork apparatus is being assembled, or alternatively that an inhibitor of replication is produced (following initiation), which then has to be diluted out. The latter would be negative control, the former positive control. There is, however, no direct evidence yet for either model.

Pontecorvo: Plainly, if we want to derive useful ideas from this work on bacteria for animal cells, we have to make comparisons with the nuclear membrane and not the cell membrane of higher cells. Is anything known about possible analogies or similarities between the plasma membrane in bacteria and the nuclear membrane in animal cells?

Dulbecco: Chromosomes in eukaryotes are attached to the nuclear membrane: this is an analogy.

Warren: The closest comparison would be between bacteria and mitochondria: both have circular DNA, and it may well be that they reproduce in the same way.

Holland: The bacterial plasma membrane has to serve several different functions; some of these are parallel to mitochondrial functions and some are equivalent to nuclear ones. In some respects therefore the bacterial membrane is likely to prove more complex than the eukaryote membrane. Nevertheless current studies of a wide variety of mammalian and bacterial membranes indicate that in all cases the majority of the membrane proteins can be accounted for by a relatively small number (20–30) of distinct polypeptides. Of course, meaningful comparisons must await the functional analysis of membrane proteins.

In reply to Professor Pontecorvo's question, I would like to mention one system in *E. coli* which may be relevant to the study of membrane function in eukaryotes. Frankel and co-workers (1968) have recently reported that upon infection of *E. coli* with bacteriophage T4, several changes in the content of membrane proteins can be detected. This system may therefore be analogous to that of the transformed cell in which Dr Dulbecco has suggested that viral-specific proteins are incorporated into the cell membrane. Further analysis of T4 infection should therefore provide a generally applicable model system for the study of the mechanism and functional significance of virus-directed changes in membrane proteins.

REFERENCES

BIRDSELL, D. C. and COTA-ROBLES, E. H. (1967) *J. Bact.* **93**, 427–437.
FRANKEL, F. R., MAJUMDAR, C., WEINTRAUB, S. and FRANKEL, D. M. (1968) *Cold Spring Harb. Symp. quant. Biol.* **33**, 495.
FRIEDMAN, R. M. (1967) *Science* **156**, 1760.
LEVY, H. B., BARON, S. and BUCKLER, C. E. (1969) In *The Biochemistry of Viruses*, p. 579, ed. Levy, H. B. New York: Marcel Dekker.
PIERUCCI, O. and HELMSTETTER, C. E. (1969) *Fedn Proc. Fedn Am. Socs exp. Biol.* **28**, 1755.
ROSENBERG, B. H., CAVALIERI, L. F. and UNGERS, G. (1969) *Proc. natn. Acad. Sci. U.S.A.* **63**, 1410.
SCHNAITMAN, C. A. (1970) *J. Bact.* **104**, 882.

GENERAL DISCUSSION

Stoker: In this meeting we have been repeatedly faced with the problem of whether we are dealing with related or independent phenomena, and this makes it hard to draw everything together. Perhaps we should concentrate on those relationships which we can identify, and which may affect the way we think about the various systems. The phenomenon which started the whole business of growth regulation, and which is still unexplained, is contact inhibition, or topoinhibition—the failure to grow in dense cultures. Some of us would like to be clearer about what the relationship of the cells is when we talk about dense cultures. How much contact is there between cells? Dr Rubin's model suggests that surface pH is important, and he can explain growth with a simple view of cells touching one another as though they were boxes of bricks, but is it really like this? What are the relationships of cells to one another in dense cultures, both normal and transformed? I'm not convinced that transformed cells ever form the same sort of dense cultures that normal cells form, anyway.

Dulbecco: I would like to challenge Dr Rubin, because he has presented experimental data, such as accurate measurements of pH, but no evidence that local changes in pH have anything to do with growth control in cultures. I would suggest the following experiment. If you have a culture with a pH higher than the optimum where it damages the cells and you make a wound, the migrating cells should have a higher surface pH than the cells in the medium and therefore should be damaged even more. An experiment of this kind, carried out at a pH just beyond the optimum, would be a better test of the local role of pH than the usual wound experiment. So far the conditions have been such that cells migrating into the wound would be in a *better* situation than those in the layer, but this can also be for other reasons. Beyond the maximum pH you should reverse this.

Rubin: I can give you a partial answer to that problem. Although we haven't looked at cells in a "wound" under high pH, we have compared sparse and dense cultures and find the former are far more vulnerable to high pH. I should be very happy to consider this a confirmation of the local pH model, but other explanations are possible. I am told that Harry Eagle finds that a number of cell lines escape density dependent inhibition if maintained at a relatively high, optimal pH.

Dr Dulbecco has fielded me an experiment; I would like to have him also field me a theory. If it turns out that all these cell lines continue to grow at maximal rate at high pH, what kind of explanation would you offer aside from proton inhibition?

Dulbecco: If you have an interaction between receptors, these interactions must involve chemical groups, and there may be ionized groups, so it would depend on the pK of these groups, and pH. I don't doubt that pH has an important role; but what a role! Think of an enzyme—for instance, lactic acid dehydrogenase. At certain pH's it doesn't work; but at other pH's the reaction works. You don't therefore say that the dehydrogenation of lactic acid is due to pH; certainly pH is required, but it is not the basic factor.

Rubin: The point is that there is a shift in the pH optimum or the pH dependency of the cells, depending on whether they are sparse or crowded, so that at the same bulk pH a crowded culture does not grow and a sparse culture does. Then by simply raising the pH by half a unit the crowded culture will grow at at least as good a rate. By the principle of parsimony we want to try to explain this phenomenon by a single mechanism. The simplest explanation, and one that does not go beyond the experimental variables, is that pH is responsible for this difference. One reasonable place to begin is to say that pH is lower at the surface of crowded than sparse cells. Inhibition of growth by low pH might work by many mechanisms: it could work through an effect on serum binding; it could work by altering membrane structure; it could work on enzyme activity. There are a great variety of possibilities, all of which should be studied as a function of a new growth parameter, namely pH.

Shodell: If, as you suggested before, more is involved in this phenomenon than simply the distance between cells, then pH is a new parameter of study only if there is also some autonomous control or cyclic change of the cell surface pH. If not, then pH may only be masking a variety of other effects. One such possibility that could be envisaged is the differential binding of active serum molecules to the cell surface as a function of pH. Thus the device of decreasing the bulk medium pH may be functionally identical to putting cells into serum-free medium.

Rubin: The observation is that crowded cells are different from sparse cells as a function of pH. You don't have to invoke cyclic changes in surface pH or anything like that. You have that simple observation to explain: whether it acts through serum binding, membrane structure or enzyme activity doesn't make much difference at this point.

Weiss: Really we want to know whether the medium *is* the message, particularly with topoinhibition. We have heard about inhibitors and serum factors; we know about density dependence, which must be a mixture of the effects of topoinhibition and serum factors. We have heard about mutants of polyoma virus that enable us to distinguish between dependence on some factors in serum and topoinhibition. Cell transformation by RNA tumour viruses may be different from transformation by DNA tumour viruses. If topoinhibited cells are infected with an RNA tumour virus, the virus is unable to overcome the topoinhibition and transformation is suppressed. Transformation of topoinhibited 3T3 cells by SV40 is similarly suppressed. However, if cells that are not proliferating because of serum depletion are infected with a DNA tumour virus, the virus can initiate the cell cycle and transform the cell. This does not happen after infection with RNA tumour viruses, although we found that the cells will transform when plated in soft agar suspension.

Dulbecco: Polyoma virus releases the cells from whatever inhibition exists in acute infection, both in the layer and in the wound. In transformation there is an additional requirement, namely integration, and maybe integration is the limiting factor which determines whether transformation (of a stable kind) will result under different conditions.

Weiss: That is a rather dogmatic statement! Integration may be important for the maintenance of the transformed state, but what about abortive transformation? There is a transient change in growth and cell behaviour; does it occur in depleted, agamma medium?

Stoker: We cannot say whether topoinhibition is affected in abortive transformation with BHK cells; perhaps with 3T3 you can?

Todaro: I'm not sure; we haven't done it. One reason we haven't is that it becomes clear that the contact-inhibited situation *is* more complicated, because you are dealing with both the effect of topoinhibition and the depletion of essential serum factors. We thought it would be simpler to study the serum factor-dependent situation, under conditions where cell-to-cell contact is not a factor and where all the cells will grow if they are given serum. In that situation, Dr Weiss is perfectly correct. DNA viruses differ in a fundamental way from RNA viruses in that they will somehow substitute for the serum factor; either SV40 or polyoma does this. RNA viruses will not, at least in our hands, directly induce the cells to resume division.

Rubin: We loosely use the terms "contact" and "monolayer", but a lot of what look like fairly well-defined monolayers turn out on closer examination to consist of areas of cells with overlapping cytoplasm and even of overlapping nuclei which would have to be considered at least in part multilayered. Can we get some indication of the extent of contact between cells in some of the conditions we are studying? What kind of interaction occurs between the bottom surface of the cell and the dish, and how does this compare with the interaction between cells in terms of closeness of apposition? Michael Stoker remarked that so-called mono-layers of even 3T3 cells may be multilayered in places and I can vouch for this in cultured chicken cells. I am convinced that simple contact is not enough for density dependent inhibition; there has to be a fairly extensive interaction between a large fraction of the cells before some effect is produced on growth of the culture.

Stoker: Following that, does this degree of contact ever occur with transformed cells?

Abercrombie: Electron micrographs show that when fibroblasts are close together there is a great deal of overlapping of very thin cytoplasmic sheets; these can be in three or four layers in what appears to be a monolayer. These overlapping cytoplasmic sheets are often extremely closely applied to each other; the gap between the cells over large areas is the order of 10–20 nm (100–200 Å), a smaller gap than that between most of the cell and the solid substrate on which it is growing.

Rubin: Can you estimate what fraction of the surface of cells is in contact with other cells?

Abercrombie: I don't think I can; I would say that this varies continuously as you increase the population. I don't believe the word "confluence" means anything much.

Weiss: Scanning electron microscopy is a simple way of visualizing cell contacts. However, it is by no means so simple to measure the amount of cell surface in contact with other cells (see Figs. 1 and 2).

Elsdale: The contact situation as such may not be the most interesting point. We are more interested in the sorts of messages that are being passed between cells, than in the physical area of contact.

Warren: But how can you separate the two aspects of the same pheno-menon? One may be the consequence of the other.

Rubin: The diameter of a 3T3 cell is about 18 or 19 μm; the average diameter of our chicken fibroblasts after they increase in size by about three-fold, which they do the day after explantation from the embryo, is about 13 μm. In the presence of 1 per cent serum or less we get a fairly sharp decrease in growth rate when the chicken cells reach about one and a half million cells. With 3T3 cells in 5 cm dishes Dr Todaro's saturation density is about a million cells. Therefore in terms of the area of the dish covered, 3T3 cells slow down at about the same point as chick embryo cells do. This is roughly the place that confluency is reached. Confluency doesn't have any exact meaning, as Michael Abercrombie indicated, because if you look carefully, even with light microscopy there is extensive overlapping

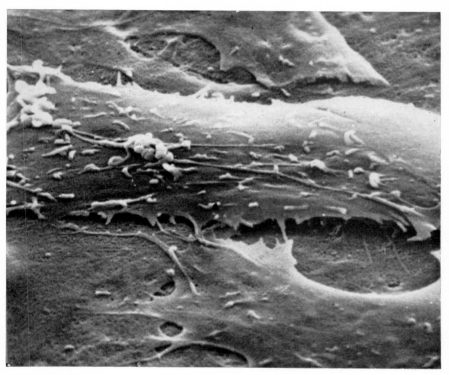

FIG. 1. (Weiss). Scanning electron micrograph of normal rat fibroblasts. Freeze dried from absolute ethanol. Tilt angle, 40°. × 10200 (A. Boyde, P. Veselý and R. Weiss.)

[*To face page 250*

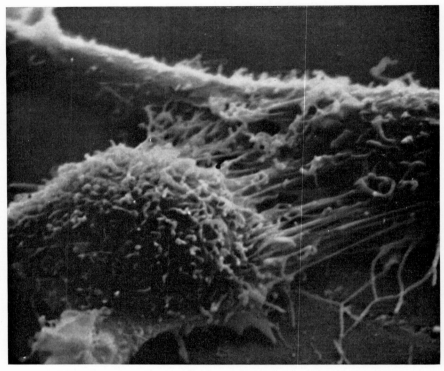

Fig. 2. (Weiss). Scanning electron micrograph of mouse sarcoma cells
(S180). Freeze dried from 70 per cent ethanol. Tilt angle, 50°. × 10600.
(A. Boyde, P. Veselý and R. Weiss.)

among the cells. I wonder to what extent Dr Todaro has seen overlapping with a million 3T3 cells in a culture?

Todaro: There is considerable cytoplasmic overlapping. The cells have long processes that run under and over adjacent cells, but in a "good" culture of 3T3 there is no nuclear overlapping.

Rubin: It's the cytoplasmic area that counts in surface interaction.

Weiss: But there is something significant here. There is a lot of cytoplasmic overlapping but there is some monolayering control mechanism because the overlapping is only on the borders.

Abercrombie: One ought to consider epithelial cultures where the situation as seen by electron microscopy is simpler than with fibroblasts; there is substantially no overlapping, though the cell border between the adjacent cells is convoluted and so is difficult to measure accurately. The surface area of an epithelial cell becomes smaller as you increase the population and this genuinely represents an increased contact with neighbouring cells; the cells become more columnar. There is a very clear positive correlation between the growth rate of such epithelia and the surface area of the cell.

Stoker: Are there any transformed epithelial cells?

Rubin: The Rous virus will transform kidney epithelium and pigmented retinal epithelial cells.

Clarke: In sparse layer cultures BHK21 and 3T3 cells resemble each other in gross morphology. As they approach confluence BHK21 cells become elongated, spindle-shaped and oriented, accommodated to a residual (minimal?) contact with the surface of the dish. 3T3 cells in the same situation spread more and become epithelioid. If this difference results from the need of 3T3 cells for a greater surface in contact with the dish as the serum concentration decreases, then even in sparse culture one might expect 3T3 cells to spread over a greater area of the dish in reduced concentrations of serum. Does this occur?

Todaro: They do appear to spread more when they are in low serum medium or "depleted" medium.

Bard: We have an observation on overlapping that is relevant here. Our fibroblasts only overlap if they can lay down collagen. If we add collagenase to the medium they don't overlap. The amount of collagen that the cells produce, if you assume it to be in the form of fibrils 100 nm (1000 Å) wide, covers only about 2 per cent of the cell surface, and this is enough to overcome contact inhibition of movement.

Stoker: To return to the difficulty of separating the effect of serum, which does after all overcome topoinhibition in most cells, and the topoinhibition itself, Dr Dulbecco has produced evidence which is supported now by other data that these are two distinct phenomena. The only cell system where they can really be distinguished yet is the BSC-1 cells.

Dulbecco: The use of a temperature-sensitive mutant of polyoma virus distinguishes between serum requirement and topoinhibition. The ts3-transformed cells at high temperature are topoinhibited but they don't require serum for DNA synthesis.

Stoker: This may then be the type of cells in which to look for the pure topoinhibition effect.

Todaro: Dr Holley suggested that perhaps everything could be explained in terms of serum factors; I wonder if we should consider that contact inhibition (or topoinhibition) really represents a change which for one reason or another increases the requirement for, or the threshold for, a serum factor? Perhaps there is a serum factor that fibroblasts need that has a hormone-like quality in much the same way as, for example, ACTH stimulates adrenal cortical cells. If we suggest this, it is a positive control; serum factor is produced somewhere in the body and is required for division of fibroblasts. Transformed fibroblasts (sarcoma cells) continue to divide in the body because they have a reduced requirement for serum factor. Is there any evidence against this theory?

Holley: I don't know any evidence against it that convinces me yet, although I think there is more than one factor required by 3T3 cells. For 3T3, I think that all the various parameters, including topoinhibition, can be overcome by one or another factor in serum; whether the effect of serum factors is primary or secondary is still uncertain.

Todaro: One could say that what the virus does is decrease the requirement for this factor. This would be consistent with most of the tissue culture evidence, and it might even explain *in vivo* evidence (Green and Todaro 1967). Cell division in the body may be controlled by a limiting concentration of serum factor and occasionally a mutant cell, maybe a virally produced mutant cell, gets below that threshold and is therefore able to grow. This would be an extreme form of the theory.

Dulbecco: The differentiation is very clear for epithelial cells like BSC-1 kidney cells, because they have high topoinhibition but don't require serum for DNA synthesis. It is very hard there to say that serum requirement plays the main role in controlling cell multiplication. My feeling is that there are two parameters we can measure, topoinhibition and serum requirements, and we should not argue whether they may be the same; we only need to find out how we are going to learn whether they are the same. It seems to me that measuring cellular overlap and so on will not take us far. The only approach is to try to isolate mutant cells which are changed in one or the other parameter.

Weiss: It would be very nice if it were true that the mutant will have a direct effect on topoinhibition and that topoinhibition isn't the result of a long series of pleiotropic effects.

Dulbecco: One must try it. You might be able to isolate it very easily; we don't know.

Bard: "Topoinhibition" is a descriptive term only. I would like to ask what are the social properties of cells which are relevant? If you put a million 3T3 cells in a 2 inch dish and add medium, a few per cent divide. If you put them in a 9 cm dish and add the same quantity of medium, they will all divide. Therefore when these cells are being crowded there must be some property which is built up and which the cells recognize as inhibiting. If I can briefly summarize what I have learned at this meeting, three possible crowding mechanisms have been put forward. The first is Harry Rubin's theory, the pH effect, which can be proved or disproved. The second comes from Dr Holland's paper, which suggests that when cells are crowded their surfaces are touching and (consequent) surface alterations could cause effects within the cell. The third possibility, that Tom Elsdale and I put forward, is that if molecules which inhibit the entry of cells into S phase have an equilibrium concentration in the culture, this will increase as the cell density increases. Essentially, this is because the higher the density, the lower is the proportion of total cell surface in contact with the medium. Can anyone think of other suitable mechanisms, preferably ones that can be tested, which can give us insights into the growth regulation system and possibly some idea of where to look for mutants?

Dulbecco: I think it very possible that there is a true effect of contact. However, what is the effective contact? It may not be identical to that detected by electrical contacts, tight junctions observable with the electron microscope and so on. It may be relevant to this problem, as said before, that one must distinguish between the transmitter of an inhibiting influence and the receiver. There could be receptors of contact inhibition on cells and these receptors may only be inhibited if they come in contact with the proper type of surface, not necessarily by just any contact. Inhibition may be very specific. If that were the case, one might isolate the transmitters of contact inhibition present on the cell surface and produce with them contact inhibition of suitable cells. This is a possible experimental approach.

Mitchison: I would like to say something here about a rather different but relevant topic: cell contact and especially matrix formation in the stimulation of lymphocytes by antigen. I raise this because it is relevant and also because it provides a technical trick that may be useful in the future. The notion is that a cell which is the precursor of an antibody-forming cell, namely a lymphocyte, has an area of immunoglobulin receptors on its surface which function as binding sites for antigen. One starts with the elementary notion that these receptors have an affinity for antigen which can be expressed in molar terms; and when the concentration of antigen in the vicinity of these receptors rises to the molar concentration corresponding to the affinity of receptors, binding takes place and the cell is then triggered into antibody formation. Although this is true in a general way, it doesn't give us much idea of what is really happening, because the concentration of antigen near these cell receptors is always much higher than it is in the medium in which the cell is growing, or in the whole mouse. The molar concentration in the whole medium is normally much lower than the average affinity of the receptors. Nevertheless, triggering occurs, so there must be local antigen concentrating mechanisms, perhaps operating by matrix formation. What is interesting is that the idea of a matrix provides a common pathway through which a lot of different mechanisms operate. When the subunits (monomers) of a polymeric protein like flagellin are polymerized, they are known to become more potent in stimulating cells antigenically. The theory is that if binding occurs initially at one point of the cell surface it raises the local concentration of antigen near the other determinants, so that multi-point binding occurs. This is why a matrix is an efficient method of locally concentrating antigen. Antigen can be presented to the receptors by auxiliary cells: a monomeric antigen is presented by a cell, either non-specifically in the case of a macrophage or by helper lymphocytes having specific receptors for the antigen. This is a reasonable interpretation of the interaction between thymus and bone marrow cells in antibody production, which has been the subject of so much interest in immunological circles for the last year or two; it is basically a matter of presenting a matrix of determinants.

Recently M. Feldman and E. Diener in Melbourne (personal communication) have obtained rather decisive evidence in favour of this theory of matrix formation in antibody production. They take a monomer, which again happens to be flagellin, and build a matrix artificially by using antibody directed against it. This matrix focuses antigen on the receptor spots on the antibody-forming cell precursor. They find that the concentration of the monomer and of the antibody used in making the matrix have to be roughly equivalent, and that may entail diluting the antiserum one million-fold in order to build an appropriate matrix.

In these kinds of cellular interactions, matrix formation seems to be an important step. If one is trying to mimic that kind of effect by using soluble extracts, the trick of throwing in matrix-building antibody is something worth bearing in mind.

Stoker: The need to isolate cell mutants in order to analyse the different phenomena has been stressed and as far as I know, nobody has isolated any membrane mutants affecting growth, but there is another sort of cell variant which can grow in the absence of serum. Dr Shodell has recently been working on one of these variants.

Shodell: This probably is not yet very relevant to topoinhibition and cell contacts. We are looking at the nature, under logarithmic growth conditions, of the requirement for serum in unadapted cells, using L cells which have been adapted by Dr S. Franks and Dr P. Riddle to grow in the complete absence of any added macromolecule. We wanted to see whether this adaptation was due to the autonomous production on the part of these L cells of whatever was originally supplied through the addition of serum. We assay this "factor" by seeing whether the medium of these adapted cells contains something which causes mitosis and growth of unadapted cells, and with the proper production and assay conditions we do find this activity. The standard assays are done in BHK21 C13 cells but the activity is also effective on chick secondary fibroblasts, secondary mouse fibroblasts and secondary hamster fibroblasts. We have fractionated it to a certain extent; the activity resides in molecules larger than 25 000 molecular weight, but it does not sediment at 100 000 **g** for 3 hours. The activity is possibly not a single component; two activities can be separated, one which seems to affect the incorporation of tritiated thymidine into unadapted cells in serum-free conditions, and the combined factors which together give the mitotic activity. There are probably at least two and possibly more than two components involved.

Montagnier: Can you propagate BHK21 cells in this type of medium indefinitely?

Shodell: That is something of a problem. The BHK cultures in the presence of the L cell factors and with no other added macromolecules will, when fixed up to four days later, show a good mitotic index. However, the total cell increment over this period is never as great as this mitotic rate would indicate. This is because not all requirements for continued cell survival in the absence of serum have been supplied by the L cell, although everything necessary for continued mitosis is apparently present. If these assays are done in the presence of the agamma serum described by Dr Todaro it is possible to obtain complete propagation of the BHK cells.

Holley: Our serum activity doesn't appear to adsorb on Sephadex at neutral pH, but it may well adsorb at low pH, and once it is purified it appears to adsorb at neutral pH.

Do you know what the concentration of protein is—how many times more active than serum it is, per mg of protein?

Shodell: It can vary according to the assay used and has generally between

50 and 150 times greater specific activity than serum in the range of 0–2 per cent. I have not yet, however, been able to achieve activities that are the equivalent of more than 5 per cent serum.

Pontecorvo: In this meeting, I have been surprised to see how rarely genetic tools are used in this field, and especially the absence of a basic approach in the use of genetic tools; that is, if you have a difference in phenotype which you believe is produced by a genetic change (in this case the oncogenic transformation which is produced either by a virus or by other means), you analyse this difference by seeing what will make one phenotype turn into the other, in this case what makes non-transformed cells turn into transformed ones or *vice versa*. A classical tool in this field is not to chase after protein, because we know that if there is a change in the genetic information there will be a change in protein at some level, but to look for the product of the activity of that protein. This is a standard genetic technique and it is probably easier than to find the protein.

Weiss: May I comment on genes and transformation by means of a cautionary tale? Let's think of a very large DNA tumour virus called a sperm. It has a restricted host range, only infecting large, non-replicating cells called eggs, which are inhibited in the prophase of an unusual cell cycle. The immediate consequences of infection are rapid changes in membrane permeability to ions and amino acids, and immunity to super-infection. Following these cell surface changes, the mitotic inhibition is released and repeated cell divisions occur. Viral genes are later expressed in the clone of cells whose growth the virus has initiated (for example, paternal antigens), and many mutant forms have been identified.

What I am saying is that the immediate responses of the egg to "infection", that is, the changes in cell surface properties and the initiation of cell division, are "elective", although the stimulus is apparently "instructive". We should bear in mind that cells infected with real tumour viruses have a repertoire of responses that may be determined by the type or state of the host cell as much as by the virus. For instance, the extent of papovavirus gene expression, leading to transformation or to lysis, is controlled by the cell; leukaemia viruses transform haemopoietic or lymphopoietic cells but not fibroblasts, although they replicate in fibroblasts; phenotypic reversion caused by BUdR or by chromosome imbalance is another example. While viral genetics is undoubtedly a most powerful tool in the analysis of cell transformation, it is also important to investigate the competence of cells to be transformed. I am trying to put forward an epigenetic view of transformation.

Montagnier: The main problem seems to me to be not the question of what the signals are which induce a change in the cell surface, but how they are transmitted from the cell surface into the cell, to the division apparatus. I would like to suggest that there may be some correlations between the sites at the cell surface and sites at the nuclear membrane. These nuclear membrane sites would be the motor sites for inducing replication of DNA and perhaps transcription; the sites on the cell surface could be the sensor sites. It should be interesting to see if there is any chemical or structural analogy between the two types of sites; perhaps the same products synthesized in the cytoplasm are going to the nuclear membrane and to the cell surface? This might be investigated by immunological techniques, by showing that there are common antigens present on the cell surface and on the nuclear membrane.

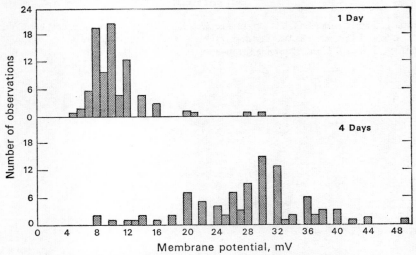

Fig. 1 (Rubin). Comparison of the membrane potential of dense and crowded cells. (P. O'Lague and H. Rubin, unpublished observations.)

Rubin: Something that hasn't been discussed is connected with the change in permeability upon fertilization referred to by Dr Weiss. This change is apparently immediately accompanied by a marked change in membrane potential of the fertilized egg; it can go from negative to positive very rapidly (Maéno 1959). We find about a three-fold difference in membrane potential between crowded cells and sparse cells; crowded cells have an average membrane potential of about −30 mV, but in uncrowded cells it is about −10 mV (see Fig. 1). This indicates that there are significant differences in the ionic composition and permeability of the cells, and probably in ratios of ions like potassium and sodium which contribute to the membrane potential. Signals have just been mentioned again, with the implication that they will be specific macromolecular signals. But there are more general signals, namely the ionic content of cells, which determines the rates of many reactions in the cell. I doubt that it is very useful to retain the term "signal" for such effects.

Pitts: We have discussed, in this symposium, a wide variety of phenomena occurring in a wide variety of cell lines and variants. I would like to see a correlation of all these variables to find out which phenotypes vary together and which phenotypes can be varied separately, to see which are related and which are not. This might be done by constructing a table relating the data on the various phenomena and cell types. (See Appendix, pp. 261–266.)

REFERENCES

GREEN, H. and TODARO, G. J. (1967) In *The Molecular Biology of Viruses*, p. 667, ed. Colter, J. S. and Paranchych, W. London and New York: Academic Press.
MAÉNO, T. (1959) *J. gen. Physiol.* **43**, 139–157.

CHAIRMAN'S CLOSING REMARKS

M. G. P. Stoker

It is now time to end the symposium, but I am not going to sum up our entire discussion.

The role of the plasma membrane in the initiation of DNA synthesis is obviously central to the whole problem we are considering. Dr Montagnier has raised some interesting points about the relationship between the plasma membrane and the nuclear membrane. Presumably changes in the plasma membrane which may be related to initiation of DNA synthesis could merely be reflexions of changes in the nuclear membrane, even though, as we know from Dr Warren, there is no direct connexion between the nuclear membrane and plasma membrane. Nevertheless, some change in nuclear membrane, perhaps insertion of a protein or a change in configuration, might be reflected in other membranes including the plasma membrane.

There are rather more interesting relationships; for example, the required control referred to by Dr Bürk, which rules that DNA should not start replicating until after cleavage of the cell, and which implies that information must be transmitted from the cell membrane to the nucleus. Also since the surface is the receptor of information about the environment, we may believe that the plasma membrane must anyway be involved in growth regulation, either directly, or through transfer of incoming molecules.

Our present difficulty arises from the multiplicity of phenomena involving growth factors: contact, surface sites and so on. Are they related in series, for example, or in parallel? One powerful method for determining the interrelationships, as Dr Pontecorvo and Dr Dulbecco have pointed out, is by genetics. It is already clear that even by isolation of one mutant of a small DNA virus, a great deal of important new information is available. An obvious task for the future will be the isolation of cell mutants, and particularly membrane mutants which affect cell growth.

Meanwhile the characterization of factors which are concerned in cell regulation will also be important. Dr Pontecorvo has asked why we investigate proteins rather than the products of the activity of proteins.

However, the active substances which have been isolated, except for the inhibitors that Dr Cunningham described, have mainly turned out to be proteins, and since we expect many regulation systems to be specific, we may end up with proteins anyway as the main objects for study.

It remains for me to thank you all for participating with energy and enthusiasm.

APPENDIX

GROWTH PROPERTIES OF NORMAL AND TRANSFORMED CELLS

John D. Pitts

Department of Biochemistry, University of Glasgow

	1 Topoinhibition	2 Saturation density	3 DNA synthesis in serum-free medium	4 Cell division in serum-free medium	5 Cell movement in serum-free medium	6 Growth inhibition by confluent 3T3 or untransformed parental cells	7 Uridine and P_i transport dependent on cell density and serum	8 Growth in agar suspension	9 Growth in agarose suspension
(a) 3T3	++++	0.5-2	-	-	-		++++	-	-
(b) 3T12	-	8-12		+++					
(c) SV3T3	-	4-25	+++	++++	+++	-		++++	
(d) SV FL101	±	1-2	+++	++++	+++	+++			
(e) SV40+3T3			+++	++					
(f) Py3T3	-	4-12	+++	++++	+++		-	++++	
(g) Py+3T3			+++		+++				
(h) BHK	+++	5-10	-		-	++++		-	++++
(i) PyBHK	-	20	+++	++++	+++	++++		++++	++++
(j) Ts3-BHK	++		+++		++			++++	
(k) Py+BHK			+++		+++			++	
(l) RSV(Bryan)-BHK		7				-		++++	++++
(m) IS-3		7						-	-
(n) Nil		2	-	-				-	++++
(o) BSC-1	++++		++++		++++				
(p) SV40+BSC-1									
(q) ME	+++	5	-	-	-			-	-
(r) Py+ME			++++						
(s) MK	++++		++++		++++				
(t) chick	+++		±	±	±	++++	++++	-	-
(u) RSV-chick						-		++++	++++
(v) HeLu				-				-	-
(w) LWF	+++	2-4	-		±	++++		-	-
(x) LW13	+	>4	±		+	+		±	±
(y) RSV-LW13	-	>5	±		+	-		++++	++++

Notes

Symbols used in table: −, negative; +, ++, +++, ++++, increasingly positive; ±, variable data; —negative or slightly positive; space, no available data.

Cell lines maintained in different laboratories may develop different properties (e.g. see note *13 o* below). The data in this table have been accumulated from several laboratories and possible variations of this nature should be kept in mind.

3v, 4v. If pH is maintained carefully at 7·5–7·6, chick cells synthesize DNA and divide in serum-free medium. The division rate is about half the maximum observed when serum is added. (Rubin, unpublished.)

10		11		12	13	14	15	16		17	18
Growth in agarose plus polyanions		Minimum pH without growth inhibition									
(A) Plus adenine	(B) Minus adenine	(A) Sparse cells	(B) Dense cells	Tumour formation (TD_{50})	Agglutination by plant agglutinins	Low-resistance junctions	Intercellular exchange of metabolites	(A) Parallel monolayer	(B) Orthogonal multilayer	Collagen formation	Karyotype (mean modal chromosome number)
					−	++++	++++	−		++++	73
				> 10⁴	+++					++++	
					++++	++++		−		++	74
					−						118
					+++						
					++++					++	70
					+++						
−	−			>10⁵	±	++++	++++	++++	++++		44
++++	++++			<10	++++	++++	+++	−			44
					±						
								−			
++++	−			<10			++++	−		−	44
				<10				++++	++++	±	
−	−			>10⁵				++++	++++	±	44
					±		++				53
							++++				
							++++				
					−						
		6·9	7·6		−	++++	++++	++++			78
					+++	++++		−			78
						++++	++++	++++	++++	++++	46
				>10⁵		++++		++++	++++		
				>10³				++++	++		
				<10				−	−		

8e. Colonies are formed with addition of 30 per cent foetal calf serum. (Stoker, unpublished.)

8h. Colonies are formed with addition of 30 per cent calf serum. (Stoker, unpublished.)

12v, 12w. RSV-chick cells form tumours in chickens while chick cells do not. (Hanafusa, Hanafusa and Rubin 1964.)

13 0. BSC-1 cells from two different laboratories behave differently after infection with SV40. BSC-1 (from R. Pollack) shows no induction of host DNA synthesis and no exposure of agglutination sites (however, infection of these cells with adenovirus, type 5, leads to induction of host DNA synthesis and exposure of agglutination sites). BSC-1 (from T. Benjamin) shows induction of host DNA synthesis and exposure of agglutination sites. (Sheppard, Levine and Burger 1971.)

CELLS

(a) *3T3*. Mouse line (fibroblast-like) with low saturation density. (Todaro and Green 1963.)

(b) *3T12*. Mouse line (fibroblast-like) related to 3T3 but with higher saturation density. (Todaro and Green 1963.)

(c) *SV3T3*. SV40 virus-transformed 3T3. (Todaro, Habel and Green 1965.)

(d) *SV FL101*. A "flat" variant of SV3T3. (Pollack, Green and Todaro 1968.)

(e) *SV40+3T3*. SV40 virus-infected 3T3 (infection leads to abortive transformation and transformation).

(f) *Py3T3*. Polyoma virus-transformed 3T3. (Todaro, Habel and Green 1965.)

(g) *Py+3T3*. Polyoma virus-infected 3T3 (lytic infection).

(h) *BHK*. Hamster line, BHK21/13 (fibroblast-like). (Macpherson and Stoker 1962.)

(i) *PyBHK*. Polyoma virus-transformed BHK.

(j) *Ts3-BHK*. BHK transformed with a temperature-sensitive mutant of polyoma virus, ts3 (Dulbecco 1971) at non-permissive temperature.

(k) *Py+BHK*. Polyoma virus-infected BHK (infection leads to abortive transformation and transformation).

(l) *RSV (Bryan)-BHK*. Rous sarcoma virus (Bryan strain)-transformed BHK.

(m) *IS-3*. Spontaneously derived, highly tumorigenic derivative of BHK21/13. (Gottlieb-Stematsky and Shilo 1964.)

(n) *Nil*. Hamster line, Nil2E (fibroblast-like) with low saturation density. (Cloned, McAllister and Macpherson 1968. Original isolation, Diamond 1967.)

(o) *BSC-1*. African green monkey kidney line (epithelial-like). (Hopps *et al.* 1963.)

(p) *SV40+BSC-1*. SV40 virus-infected BSC-1 (lytic infection).

(q) *ME*. Mouse embryo primary cultures (fibroblast-like).

(r) *Py+ME*. Polyoma virus-infected ME (lytic infection).

(s) *MK*. Baby mouse kidney primary cultures (epithelial-like).

(t) *Chick*. Chick embryo primary cultures (fibroblast-like).

(u) *RSV-chick*. Rous sarcoma virus-transformed chick.

(v) *HeLu*. Human embryonic lung primary cultures (fibroblast-like).

(w) *LWF*. Lewis rat embryo primary cultures (fibroblast-like). (Veselý *et al.* 1968.)

(x) *LW13*. Lewis rat line (fibroblast-like). (Veselý *et al.* 1968.)

(y) *RSV-LW13*. Rous sarcoma virus (Prague strain)-transformed LW13. (Veselý *et al.* 1968.)

PROPERTIES

(1) *Topoinhibition*. Proportion of cells incorporating [^{3}H]thymidine in freshly wounded area of cell sheet compared to proportion incorporating in undisturbed area of cell sheet. (Dulbecco 1970.)

(2) Saturation density. Final density (cells/cm$^2 \times 10^{-5}$) when cells are cultured in excess medium supplemented with 10 per cent serum. (Todaro, Habel and Green 1965.)

(3) DNA synthesis in serum-free medium. Incorporation of [^{3}H]thymidine by cells in sparse culture (eg. in wound of confluent monolayer) in serum-free defined medium. (Cf. wound serum requirement, Dulbecco 1970.)

(4) Cell division in serum-free medium. Persistent cell division when cultured in serum-free defined medium.

(5) Cell movement in serum-free medium. Persistent cell movement when cultured in serum-free defined medium. (Dulbecco 1970).

(6) Growth inhibition by confluent 3T3 cells or untransformed parental cells. Inhibition of growth by contact with static 3T3 or untransformed parental cells. (Weiss and Njeuma 1971.)

(7) Uridine and P$_i$ transport dependent on cell density and serum. Decrease in transport after growth to confluency and rapid stimulation of transport after addition of fresh serum to confluent cells. (Cunningham and Pardee 1969.)

(8) Growth in agar suspension. Formation of large colonies from single cells suspended in nutrient agar.

(9) Growth in agarose suspension. Formation of large colonies from single cells suspended in nutrient agarose.

(10) Growth in agarose plus polyanions, + adenine and − adenine. Cell growth in agarose containing dextran sulphate in the presence and absence of adenine or adenine nucleotides. (Montagnier 1970.)

(11) Sparse cells/dense cells: minimum pH without inhibition. pH below which the growth of cells in sparse or dense cultures is inhibited. (Rubin 1971.)

(12) Tumour formation (TD$_{50}$). Number of cells required to form, on inoculation, tumours in 50 per cent young, non-irradiated animals.

(13) Agglutination by plant agglutinins. Data refer to agglutination by wheat germ agglutinin (Burger and Goldberg 1967). Similar results are obtained with phytohaemagglutinin, *Lens culinaris* agglutinin (M. M. Burger, unpublished) and Concanavalin A (Inbar and Sachs 1969).
(14) Low-resistance junctions. Formation of cell-to-cell junctions of low electrical resistance. (Potter, Furshpan and Lennox 1966.)
(15) Intercellular exchange of metabolites. Exchange of metabolites between cells in contact. (Pitts 1971*a, b*.)
(16) Parallel monolayer; orthogonal multilayer. Orientation of cells when grown to confluence in excess medium. (Elsdale and Foley 1969.)
(17) Collagen formation. (Green, Goldberg and Todaro 1966.)
(18) Karyotype (mean modal chromosome number).

DATA SOURCES

(1) *a, c, d, h, i, o, q*, Dulbecco 1970; *b*, Todaro, unpublished; *f, j, s*, Dulbecco, unpublished; *t*, Gurney 1969; *w, x, y*, Weiss and Veselý, unpublished.
(2) *a*, Todaro, Habel and Green 1965; *b, c, d, f*, Todaro, unpublished; *h, i, m*, House and Stoker 1966; *l*, Stoker, unpublished; *n*, Diamond 1967; *q*, Weiss 1970*a*; *w, x, y*, Weiss and Veselý, unpublished.
(3) *a, c, d, h, i, o, q*, Dulbecco 1970; *e, f, g, j, s*, Dulbecco, unpublished; *k*, Taylor-Papadimitriou, Stoker and Riddle 1971; *n*, Stoker, unpublished; *t*, Shodell and Rubin 1970; *w, x, y*, Weiss and Veselý, unpublished.
(4) *a*, Holley, unpublished; *b, c, d, e, f*, Todaro, unpublished; *i*, Bürk 1967; *n*, Stoker, unpublished; *q*, Pitts, unpublished; *t*, Shodell and Rubin 1970; *v*, Elsdale, unpublished.
(5) *a, h, k*, Stoker, unpublished; *c, d, f, g, i, j, o, q, s*, Dulbecco, unpublished; *t*, Rubin, unpublished; *w, x, y*, Weiss and Veselý, unpublished.
(6) *c*, Todaro and Green 1963; *d*, Pollack, Green and Todaro 1968; *h, i*, Stoker, Shearer and O'Neill 1966; *h, i, l, u*, Weiss 1970*a*; *t*, Weiss and Njeuma 1971; *w, x, y*, Weiss and Veselý, unpublished.
(7) *a, f*, Cunningham and Pardee 1969; *t*, Rubin 1970, Weber and Rubin 1971, and Weber and Edlin, unpublished.
(8) *a*, Stoker, unpublished; *c, f*, Todaro, unpublished; *h. i*, Montagnier and Macpherson 1964; *e*, Montagnier and Vigier 1967; *j*, Dulbecco, unpublished; *k*, Stoker 1969; *l*, Macpherson 1966; *m*, Macpherson 1968; *q, t*, Montagnier, unpublished; *u*, Weiss 1970*b*, Rubin, unpublished; *v*, Elsdale, unpublished; *w, x, y*, Weiss and Veselý, unpublished.
(9) *a, m*, Stoker, unpublished; *h, i, l*, Montagnier 1968; *n, q*, Montagnier, unpublished; *t, u*, Weiss, unpublished; *v*, Elsdale, unpublished; *w, x, y*, Weiss and Veselý, unpublished.
(10) *h, i, l*, Montagnier 1968; *n*, Stoker, unpublished.
(11) *t*, Rubin, unpublished.
(12) *h, m*, Macpherson 1968; *i*, Macpherson 1963; *l*, Macpherson 1966; *n*, Diamond 1967; *w, x, y*, Weiss and Veselý, unpublished.
(13) *a, c, f, t, u*, Burger 1970; *b, d*, Pollack and Burger 1969; *e, o*, Inbar and Sachs 1969 and Sheppard, Levine and Burger 1971; *g*, Benjamin and Burger 1970; *h, i*, Burger and Goldberg 1967; *j*, Eckhart, Dulbecco and Burger 1971; *s*, Burger, unpublished.
(14) *a, c, h, i*, Potter, Furshpan and Lennox 1966; *t, u*, O'Lague and Rubin, unpublished; *w*, Weiss and Veselý, unpublished.
(15) *a*, Dulbecco, unpublished; *i*, Bürk, Pitts and Subak-Sharpe 1968; *h*, Pitts 1971*b*; *l, o, q, r, t, v*, Pitts, unpublished.
(16) *a, c*, Stoker, unpublished; *h*, Macpherson and Stoker 1962; *i*, Macpherson 1963; *k*, Taylor-Papadimitriou, Stoker and Riddle 1971; *l*, Macpherson 1966; *m*, Macpherson 1968; *n*, Diamond 1967; *t, u*, Rubin, unpublished; *v*, Elsdale and Foley 1969; *w, x, y*, Weiss and Veselý, unpublished.
(17) *a, b, c, f*, Green, Goldberg and Todaro 1966; *l, m, n*, Stoker, unpublished; *v*, Elsdale and Foley 1969.
(18) *a, c, d, f*, Pollack, Wollman and Vogel 1970; *t, u*, Temin and Rubin 1958.

REFERENCES

BENJAMIN, T. L. and BURGER, M. M. (1970) *Proc. natn. Acad. Sci. U.S.A.* **67**, 929–934.

BURGER, M. M. (1970) In *Permeability and Function of Biological Membranes*, pp. 107–119, ed. Bolis, L. Amsterdam: North Holland.

BURGER, M. M. and GOLDBERG, A. R. (1967) *Proc. natn. Acad. Sci. U.S.A.* **57**, 359–366.

BÜRK, R. R. (1967) In *Growth Regulating Substances for Animal Cells in Culture*, pp. 39–50, ed. Defendi, V. and Stoker, M. Wistar Institute Symposium Monograph No. 7. Philadelphia: Wistar Press.

266 APPENDIX

Bürk, R. R., Pitts, J. D. and Subak-Sharpe, J. H. (1968) *Expl Cell Res.* **53**, 297–301.
Cunningham, D. D. and Pardee, A. B. (1969) *Proc. natn. Acad. Sci. U.S.A.* **64**, 1049–1056.
Diamond, L. (1967) *Int. J. Cancer* **2**, 143–152.
Dulbecco, R. (1970) *Nature, Lond.* **227**, 802–806.
Dulbecco, R. (1971) This volume, pp. 71–76.
Eckhart, W., Dulbecco, R. and Burger, M. M. (1971) *Proc. natn. Acad. Sci. U.S.A.* **68**, 283–286.
Elsdale, T. and Foley, R. (1969) *J. Cell Biol.* **41**, 298–311.
Gottlieb-Stematsky, T. and Shilo, R. (1964) *Virology* **22**, 314–320.
Green, H., Goldberg, B. and Todaro, G. (1966) *Nature, Lond.* **212**, 631–633.
Gurney, T. (1969) *Proc. natn. Acad. Sci. U.S.A.* **62**, 906–911.
Hanafusa, H., Hanafusa, T. and Rubin, H. (1964) *Proc. natn. Acad. Sci. U.S.A.* **51**, 41–48.
Hopps, H. E., Bernheim, B. C., Nisalak, A., Tijo, J. H. and Smadel, J. E. (1963) *J. Immunol.* **91**, 416–424.
House, W. and Stoker, M. G. P. (1966) *J. Cell Sci.* **1**, 169–173.
Inbar, M. and Sachs, L. (1969) *Proc. natn. Acad. Sci. U.S.A.* **63**, 1418–1425.
McAllister, R. and Macpherson, I. (1968) *J. gen. Virol.* **2**, 99–106.
Macpherson, I. (1963) *J. natn. Cancer Inst.* **30**, 795–815.
Macpherson, I. (1966) In *Recent Results in Cancer Research*, no. 6, pp. 1–11.
Macpherson, I. (1968) *Int. J. Cancer* **3**, 654–662.
Macpherson, I. A. and Stoker, M. G. P. (1962) *Virology* **16**, 147–151.
Montagnier, L. (1968) *C.r. hebd. Séanc. Acad. Sci., Paris, Sér. D.* **267**, 921–924.
Montagnier, L. (1970) In *Defectivité, démasquage et stimulation des virus oncogènes*, pp. 19–32. Paris: Centre National de la Recherche Scientifique.
Montagnier, L. and Macpherson, I. (1964) *C. r. hebd. Séanc. Acad. Sci., Paris, Sér. D.* **258**, 4171–4173.
Montagnier, L. and Vigier, P. (1967) *C.r. hebd. Séanc. Acad. Sci., Paris, Sér. D.* **264**, 193–196.
Pitts, J. D. (1971*a*) This volume, pp. 89–98.
Pitts, J. D. (1971*b*) In preparation.
Pollack, R. E. and Burger, M. M. (1969) *Proc. natn. Acad. Sci. U.S.A.* **62**, 1074–1076.
Pollack, R., Green, H. and Todaro, G. (1968). *Proc. natn. Acad. Sci. U.S.A.* **60**, 126–133.
Pollack, R., Wollman, S. and Vogel, A. (1970) *Nature, Lond.* **228**, 938–970.
Potter, D. D., Furshpan, E. J. and Lennox, E. S. (1966) *Proc. natn. Acad. Sci. U.S.A.* **55**, 328–336.
Rubin, H. (1970) In *Second International Symposium on Tumor Viruses*, pp. 11–17, ed. Boiron, M. Paris: Centre National de la Recherche Scientifique.
Rubin, H. (1971) This volume, pp. 127–145.
Sheppard, J. R., Levine, A. J. and Burger, M. M. (1971) In preparation.
Shodell, M. and Rubin, H. (1970) *In vitro* **6**, 66–74.
Stoker, M. (1969) *Nature, Lond.* **218**, 234–238.
Stoker, M. G. P., Shearer, M. and O'Neill, C. (1966) *J. Cell Sci.* **1**, 297–310.
Taylor-Papadimitriou, J., Stoker, M. and Riddle, P. (1971) *Int. J. Cancer* **7**, 269.
Temin, H. and Rubin, H. (1958) *Virology* **6**, 669–688.
Todaro, G. J. and Green, H. (1963) *J. Cell Biol.* **17**, 299–313.
Todaro, G. J., Habel, K. and Green, H. (1965) *Virology* **27**, 179–185.
Veselý, P., Donner, L., Cinatl, J. and Sovova, V. (1968) *Folia Biol., Praha* **14**, 457–465.
Weber, J. M. and Rubin, H. (1971) *J. cell. Physiol.* In press.
Weiss, R. A. (1970*a*) *Expl Cell Res.* **63**, 1–18.
Weiss, R. A. (1970*b*) *Int. J. Cancer* **6**, 333–345.
Weiss, R. A. and Njeuma (1971) This volume, pp. 169–184.

INDEX OF AUTHORS *

Page numbers in bold type indicate papers; other page numbers are contributions to the discussions.

* Author and subject indexes prepared by William Hill.

Printed by William Clowes & Sons Limited, London, Colchester and Beccles